Hobby Farming

FOR

DUMMIES®

by Theresa A. Husarik

Wiley Publishing, Inc.

Hobby Farming For Dummies®

Published by
Wiley Publishing, Inc.
111 River St.
Hoboken, NJ 07030-5774
www.wiley.com

Copyright © 2008

Published simultaneously in Canada

No part of this publication may be reproduced, stored in a retrieval system or transmitted in any form or by any means, electronic, mechanical, photocopying, recording, scanning or otherwise, except as permitted under Sections 107 or 108 of the 1976 United States Copyright Act, without either the prior written permission of the Publisher, or authorization through payment of the appropriate per-copy fee to the Copyright Clearance Center, 222 Rosewood Drive, Danvers, MA 01923, 978-750-8400, fax 978-646-8600. Requests to the Publisher for permission should be addressed to the Legal Department, Wiley Publishing, Inc., 10475 Crosspoint Blvd., Indianapolis, IN 46256, 317-572-3447, fax 317-572-4355, or online at http://www.wiley.com/go/permissions.

Trademarks: Wiley, the Wiley Publishing logo, For Dummies, the Dummies Man logo, A Reference for the Rest of Us!, The Dummies Way, Dummies Daily, The Fun and Easy Way, Dummies.com, and related trade dress are trademarks or registered trademarks of John Wiley & Sons, Inc. and/or its affiliates in the United States and other countries, and may not be used without written permission. All other trademarks are the property of their respective owners. Wiley Publishing, Inc. is not associated with any product or vendor mentioned in this book.

For general information on our other products and services, please contact our Customer Care Department within the U.S. at 800-762-2974, outside the U.S. at 317-572-3993, or fax 317-572-4002.

For technical support, please visit www.wiley.com/techsupport.

Wiley also publishes its books in a variety of electronic formats. Some content that appears in print may not be available in electronic books.

Library of Congress Control Number: 2008

ISBN: 978-0-470-28172-7

Manufactured in the United States of America

10 9 8 7 6 5 4 3 2 1

Hobby Farming For Dummies®

Published by
Wiley Publishing, Inc.
111 River St.
Hoboken, NJ 07030-5774
www.wiley.com

About the Author

Theresa A. Husarik is a writer, photographer, crafter, fiber person, and animal lover who lives on a small plot far away from the heart of the city. When she is not tending to her brood (which includes llamas, alpacas, angora goats, cats, dogs, peacocks, and chickens), she can usually be found either behind the computer writing something or in the craft room making something.

Dedication

To Max, for shining his light in my life if only for too short a time.

Author's Acknowledgments

Thanks to Charlie for enthusiastically sharing his knowledge of all things farming; to Todd for helping with the parts dealing with animals; to my agent, Barb Doyen, for finding this opportunity for me; and to all the editors at Wiley for helping make this book the best it can be.

Publisher's Acknowledgments

We're proud of this book; please send us your comments through our Dummies online registration form located at www.dummies.com/register/.

Some of the people who helped bring this book to market include the following:

Acquisitions, Editorial, and Media Development

Project Editors: Stephen R. Clark, Natalie Faye Harris

Acquisitions Editor: Mike Baker

Senior Copy Editor: Danielle Voirol

Editorial Program Coordinator: Erin Calligan Mooney

Technical Editor: Susan Brooks

Editorial Manager: Christine Meloy Beck

Editorial Assistants: Joe Niesen, David Lutton

Cover Photos: Mike Hill

Cartoons: Rich Tennant (www.the5thwave.com)

Composition Services

Project Coordinator: Erin Smith

Layout and Graphics: Reuben W. Davis,

Special Art: Lisa Reed

Proofreaders: John Greenough, Penny Stuart

Indexer: Beth Palmer

Special Help: Alicia South, Emily Nolan, Connie Sarros

Publishing and Editorial for Consumer Dummies

> **Diane Graves Steele,** Vice President and Publisher, Consumer Dummies
>
> **Joyce Pepple,** Acquisitions Director, Consumer Dummies
>
> **Kristin A. Cocks,** Product Development Director, Consumer Dummies
>
> **Michael Spring,** Vice President and Publisher, Travel
>
> **Kelly Regan,** Editorial Director, Travel

Publishing for Technology Dummies

> **Andy Cummings,** Vice President and Publisher, Dummies Technology/General User

Composition Services

> **Gerry Fahey,** Vice President of Production Services
>
> **Debbie Stailey,** Director of Composition Services

Contents at a Glance

Table of Contents

Part II: Down on the Farm: Getting Your Property in Order......................................63

Chapter 5: Getting Out of the House: Outbuildings and Enclosures 65

Chapter 6: Gearing Up with the Right Tools, Clothes, and Equipment 77

Introduction

So many people dream of moving into the country and away from the hustle and bustle of city life. Everyone gets tired of traffic and smog and long lines and crowds. Getting back to the simpler life so often seems like a dream that you wonder why you didn't do it sooner.

If you have preconceived notions of what farming is like today, you may be surprised. Rural families don't necessarily spend all day in overalls on the back of a tractor or in the kitchen donning an apron and baking pies. Every farming operation is different — in size, products, use of technology, and so on. Farmers fill a variety of roles (mechanic, veterinarian, carpenter, gardening expert, and accountant), and although taking on so many mini careers at once may seem daunting, you can pick up experience on the job. The topics covered in this book can get you started.

My husband and I moved to a farm from the city about ten years ago. Yes, we've had some hard times (such as dealing with blizzards and droughts and numerous power outages), but the whole experience has been so wonderful for both of us. Our animals are a daily delight — each time we go out to feed, one of us announces, "I got a kiss from Alex!" or "Oggie let me pet and hug him!" And in the summertime, when we go out to harvest, it's such a joy to cut up a tomato or zucchini and know that we made that.

If you likewise delight in an accomplishment such as growing the biggest tomato in the county or seeing the birth of a calf or making a salad using the wonderfully tasty veggies you grew yourself, then this is the life for you. If you *like* to get your hands dirty and put in some hard work, hobby farming is very likely going to be all you dreamed of and more. We learned a lot in the first few years of playing with this new toy (the farm with animals), and as I share some of my experiences with you, I hope to encourage and inform you so you're more prepared to go for your dream.

About This Book

These days, you can go to any online search engine, type in a phrase such as "basic animal care," and get tons of Web sites that'll potentially give you answers. Or you can always do your research the old-fashioned way by going to the library and poring over multiple books. But sorting through all that information can be a bit like looking for the proverbial needle in a haystack. I hope that in all my research and personal anecdotes, you find that this book is a one-stop place for all you need to know about deciding whether to take the plunge into the rural life.

Farming today is diverse, and I can't cover all the intricacies of alpaca diet and chile pepper varieties and windmill construction. But with *Hobby Farming For Dummies,* you can get the foundation for everything you need and lots of pointers to other, more-intensive resources, along with some personal stories that can help you understand what you need to succeed in this lifestyle. The info you find here, along with many of the other books and resources I recommend, can help you (and all those who jump, with reckless abandon, into this really fun and rewarding hobby) to realize this dream — or at least give you the skills to keep your head above water and your heart fluttering with joy about your decision to head this way.

The book is organized in standalone chapters. You can start with Chapter 1 and read all the way through, or you can skip to Chapter 11 just to find out what to do with animals. Each chapter is ready and waiting for the day you're interested in the topic covered. Inside each chapter are tips and information about how you can deal with the particular topic as well as directions to other resources to give you all the help you need.

Conventions Used in This Book

Although I've tried to give you a good overview of each topic selected in this book, I've also included many Web sites or books to give you a more extensive look at each topic presented. Web addresses appear in monofont.

When this book was printed, some Web addresses may have needed to break across two lines of text. If that happened, rest assured that I haven't put in any extra characters (such as hyphens) to indicate the break. So when using one of these Web addresses, just type exactly what you see in this book, pretending as though the line break didn't exist.

Farming also has its own lingo, so any new terms that I present appear in *italics,* immediately followed by a plain-English definition.

Also, all temperatures appear in degrees Fahrenheit.

What You're Not to Read

This book includes many sidebars — little gray boxes of additional information. Sidebars provide either a more extensive, scientific look at a topic or a personal anecdote on how a certain topic affected my farm and family. They can give you additional insight, but they're beyond the basics, so feel free to skip them if you have to make a feed store run or want to take a look at the used tractor you found in the classifieds.

You can also gloss over anything marked with a Technical Stuff icon, which marks info that's interesting but isn't quite need-to-know.

Foolish Assumptions

In order to make this book as helpful (and relevant and easily readable) as possible, I've presented all the ideas in a friendly, informal, conversational style — as though we were sitting out on the porch with a glass of cold lemonade and watching the clouds go by. Here are some assumptions I've made about you:

✔ You're intelligent and may know a great deal about a lot of things (perhaps in your current life you're a doctor, lawyer, business executive, or college professor), but life on the farm is new to you.

✔ You want a good idea ahead of time (before you jump in with both feet) of what this lifestyle has in store for you.

✔ You want to know how to do things right and how to choose the operations that'll work best for you. You're looking for a good reference for those times when you're starting something new.

✔ You have some basic carpentry, gardening, and animal-raising skills — or you're willing to put in the time and effort to develop them.

How This Book Is Organized

This book is divided into seven parts that group related chapters and appendixes. Here's a preview of what you can find in each part.

Part I: Hobby Farms 101

In this part, I point out the basics of life on the farm to help you decide whether rural life is a good fit for you. I discuss farming challenges, cultural differences, and utilities — such as water and sewage — that you may have to provide for yourself. I also help you evaluate properties and give you advice for fitting in with your new neighbors.

Part II: Down on the Farm: Getting Your Property in Order

In Part II, I talk about the tools that can help you develop a relationship with the land. You discover outbuildings that provide shelter and storage, machines that help you with farm labor, and power sources to keep your equipment up and running. I also talk about the property itself — water sources, hills, pastures, and so on — and how you can access and care for it.

Part III: Calling In the Critters: What's a Farm without Animals?

Whether as pets, food sources, laborers, or sources of fun, animals are a farm institution. In this part, you discover types of animals available, where you can buy your livestock, how to provide critters' basic needs, and what to do when an animal gets sick or injured.

Part IV: Choosing and Growing Plants and Crops

Plants take a starring role on many farms — from potted herbs to acres of corn, from food for your table to food for the market, and from salad trimmings to nutrient-rich pastures for your four-legged friends. In this part, I help you decide how big you want to go with your growing operation, advise you on garden plans, and give you some info on basic plant care.

Part V: Using the Fruits of Your Labor

Sure, your friends may love gifts of live chickens and bags of wool, but you don't have to stop there. Put those beautiful fruits, vegetables, and animal products to good use! Make your own cheese or yogurt, turn fiber into a winter scarf, and enrich your soil with manure and compost. In this part, I introduce some creative ideas that may help expand your hobbies beyond farming itself.

Part VI: The Part of Tens

The Part of Tens gives you some quick info in classic top-ten format. I answer ten common questions about farm life and mention some ways to have a lot of fun and enjoy rural culture.

Part VII: Appendixes

In the appendixes, I list Web sites and other books and magazines where you can go to get more in-depth information about the topics I've presented. You also find a list of terms that can help you sound like a local as you shoot the breeze with your neighbors.

Icons Used in This Book

The icons in this book help you quickly identify specific kinds of information that may be of use to you:

The Remember icon highlights important ideas for you to keep in mind to deepen your understanding of hobby farming.

This icon draws attention to points that help you make sense of hobby farming and save you some time and frustration.

Steer clear of the pitfalls flagged in the Warning paragraphs.

The Technical Stuff icon highlights info that's useful to know but isn't essential reading.

Where to Go from Here

This book isn't linear — you can start anywhere and skip around to those topics that interest you or apply to your situation. Don't worry that reading a chapter without looking at the one before it will leave you scratching your head. You can easily read Chapter 9 first and then check out what's in Chapter 5. You won't get lost, I promise.

If you're familiar with life on the farm (maybe you grew up on a farm but spent several years at a job in the city and now want to return to your roots), you may be able to skip the chapters that introduce the lifestyle. You probably already know how much work there'll be and that you'll probably have to provide your own amenities. Or if you've bought and sold houses many times, you may not be interested in the chapter about selecting your property.

But maybe you've never had farm animals before, so the chapters on animal care, medical attention, and where to get animals appeal to you. Or maybe you're the first-timer, born and raised in the city (without even a grandma or uncle with a house on the farm that you could visit). If that's your situation, you probably want to start at the beginning.

Part I
Hobby Farms 101

The 5th Wave   By Rich Tennant

"Janet and I always wanted to get away from the city, own a small farm, and grow calluses together."

In this part . . .

Oh, the joys and challenges of country life! You can enjoy beautiful scenery, fresh air galore, no long lines of traffic, and lots of quiet. But this life comes with its share of problems, too, such as droughts (or floods) and power outages. Being so far out means you have to be at least somewhat self-sufficient and be prepared for whatever you may encounter.

This part discusses some challenges that a small farmer faces, as well as some of the amenities you may have to provide for yourself as a hobby farmer. It also helps you make the distinction between a for-profit farm and a hobby farm and gives you advice on choosing your property.

Chapter 1

Heading for the Country

- -

In This Chapter

▶ Determining whether this is the life for you

▶ Deciding which operations to get into

▶ Getting ready to go

- -

So you think you want to move away from the hubbub of the city and move to the peaceful life of the country. Doing that offers so many rewarding consequences, but you also have to keep some challenges in mind before taking the plunge and buying the farm. After all, you want it to be a pleasant experience that doesn't do you in!

One of the best parts of hobby farming is the wonderful feeling of getting your hands dirty, growing a small crop of farm-fresh vegetables, and ending up with something you can actually eat and enjoy. Or perhaps you experience the joy and wonder that occurs after you care for a pair of animals who've given birth. You can proudly show off your crops and newborns to your friends and relatives, saying, "Look what I did!"

But the downsides include simple inconveniences such as being farther away from the grocery store, eateries, or even the fire department. Most farm areas are also more exposed to dangerous weather just because they're out in the open. And then you face the issues that come from drought, insect attacks, and too much work but too little time.

But with all the trade-offs, you just may find the country life is the best thing that ever happened to you and wonder why you didn't embark on that journey sooner. In this chapter, I touch on some of the pros and cons to consider and point you to other places in the book where you can find more-detailed information.

Analyzing the Lifestyle

Think about why, really, you're considering the move and the lifestyle change. Are you trying to escape the city, or do you truly enjoy being one with the earth and getting your hands dirty on a daily basis? If escape from city hassles is your motivation, just be aware that you're trading one set of hassles for another. In this section, you discover some of the pros and cons of country life.

Looking at some drawbacks

Moving from the city or suburbs to the country isn't always a smooth transition, especially if you decide to build your own place. For instance, if your homestead isn't close enough to the city's or county's services, you may need to get your own propane tank or use some other alternative power source (see Chapter 7), dig a well that can cost several thousand dollars, and set up a septic system to deal with household wastes. Besides the initial costs of building these types of systems, you have to put in time and money to maintain them. Chapters 2 and 3 go into more detail about these utilities.

Even if public services such as power, water, and sewage are available, you'll still run into some problems — power outages, road washouts, or a drought that can threaten your crops.

Ask yourself whether you're willing to deal with the trade-offs, such as not having the opera nearby but instead going to the local high school musical for your cultural entertainment. Are you willing to give up short trips to the store for clean air and a quieter life? No traffic for no movie theater around the corner? Chapter 2 discusses some other sacrifices you may not have thought about.

And consider the concept of fitting into the neighborhood. You want to do your thing without making enemies of those who are already up and running. Respect what's already going on and try not to make big changes that may cause rifts between you and your neighbors (such as deciding you don't like the odor coming from your neighbor's pig farm and attending every city council meeting to voice your opinion about it, hoping to get the operation to cease and desist). You can't always pick your neighbors, and making a big investment in land, animals, and equipment means you can get yourself into a situation where you can't easily walk away. Look and *smell* before you buy! Chapter 4 discusses some ways you can ensure you're a good neighbor.

And of course, farming is hard work. There will always be a fence to mend, an animal shelter to build, or planting and canning to tackle. Although many tasks can wait, some of them have to be done immediately. Much of this book explains what kinds of work you may be taking on.

Looking at a few rural benefits

Despite the challenges and rigors of farm life, you do get some benefits that you don't find elsewhere — quiet streets, being able to see stars and meteor showers at night, and neighbors who tend to look out for each other and lend a hand when needed.

Growing your own food and knowing just what kinds of processes and chemicals were involved (if any) is one of the joys of farm life. And homegrown fruits and veggies just taste better. After you've eaten your own fresh-from-the-garden tomatoes, you won't be satisfied with a store-bought one again.

Having animals can be a lot of fun (but a lot of work as well). You can raise animals merely for the joy they provide, or you can get them to help you in some way — by pulling equipment, giving you meat or fresh eggs, or giving you fiber that you can turn into beautiful fabric.

Envisioning Your Farm

Your motives for moving to a farm, whether you want to raise plants or animals or both, tie in with choosing a location. And if you're not going solo, a lot of what you decide depends on what the rest of the family has to say, because everyone wants some input on such a big change. The bottom line is that you need to look at a lot of interrelated issues that go into making your farm dream come true. In this section, I discuss these considerations from different angles.

Size matters: Comparing a small hobby farm to a for-profit operation

Whether you're looking at a true hobby farm as opposed to a fully for-profit farming venture clearly impacts your perspective on the shape your farm should take. _Hobby_ means doing something for the fun of it, not necessarily to make a living. Moving to a small hobby farm means you'll likely be pouring more money into it than you get out (if you get something out of it at all). And most likely, you'll be working in your "real" job and commuting into town every day. This leaves less time for your farming chores and challenges.

If you're looking to go after a for-profit venture, your decisions will be quite different. Odds are, if this is the case, the farm will be your only job, so you need to put more thought into planning, management, budgets, and marketing. If your bottom line depends on what you put into and subsequently get out of the farm, you may not be as free to buy all those fun toys that tempt you (such as a backhoe or a top-of-the-line computerized greenhouse). You'll need to be prepared to suffer monetary losses — and these can be substantial — until you're able to get the farm fully productive and then profitable.

Choosing your venture: What you want to do with your farm

Before you can decide where to locate your hobby's headquarters, you have to decide just what you want to do with that farm. The best hobby farms begin with a vision, so think carefully about your goals. Write down your answers to some of the following questions:

✔ Are you interested in having animals, and if so, which ones have captured your heart? (See Chapter 9 for some typical farm animals.) Do you want these animals as companion animals or to serve a working purpose, such as providing you with eggs or fiber or even helping out with farm chores?

✔ What kinds of plants do you want to grow, and just how much do you want to deal with? Do you want to supplement your grocery list, grow some to sell for a little bit of profit, or maintain a subsistence-level operation? Chapter 13 gives you information on growing plants, and Chapter 14 gives you some tips on getting those plants started.

✔ What are your social goals? Do you have a desire to do the farmer's market or roadside stand circuit, where you get to interact with the general public? Do you have an interest in inviting strangers to your home so *they* can have a farm vacation?

In this section, I describe some of what you can do on your farm. After you figure out what you want, you can think about what you need — in terms of people, skills, time, labor, money, natural resources, and so on — and plan how to get there.

Raising animals down on your farm

What kinds of animals you decide to bring into your family is a personal decision. Each animal and type of animal comes with unique needs, some more demanding than others. You need to think about what's right for you. Animals all need to be fed and watered daily, and some need some extra care above and beyond those basics. And even within a species, individual animals have unique personalities and quirks that can be endearing or more work than you signed up for.

Deciding which animals you want

At the very least, you want some sort of alarm animal. Most people opt for a watchdog. Especially if your nearest neighbor is a mile away, a dog can contribute to your sense of well-being and can warn you if something is amiss. Dogs are social animals, so treat them as part of the family.

Also, a couple of cats are great to have around to help keep the rodent population down. Everybody works on the farm!

Here are some other animals to consider:

- ✔ **Horses:** Horses need the most amount of work because they have to be groomed and exercised daily, but they can provide a lot of joy when they take you for a ride across the meadows or up into the mountains.

- ✔ **Fiber animals:** Animals such as sheep, alpacas, Angora goats or rabbits, and llamas have to be kept in such a way that their fur or wool is protected.

- ✔ **Meat and dairy animals:** You may keep some animals, such as chickens, cows, sheep, or goats, because you want them for the meat. Goats and cows are common milk producers. Health and nutrition is an especially big consideration for animals who produce anything you plan to consume.

- ✔ **Companion animals:** Not every typical meat animal ends up as dinner. Even cows can serve as pets.

Chapter 9 discusses some of the common and a few exotic animals who live on farms.

Understanding the care involved

If you do bring animals into your life, you're responsible for their care. That means daily attention as well as stepping in when they need more help. Even daily care can be a real chore — during a blizzard, for instance, you may need to trudge through the snow to haul water so the animals have something other than ice in their bowls. Also, animals do get sick, and when that happens, you have to deal with it or suffer the consequences. Sometimes animals have difficult births, and you have to help or else the young one may not make it. Chapter 11 gives you the basics of animal care, and Chapter 12 goes into what to do when animals need medical attention.

You may end up with an animal who just doesn't get along with the rest of the herd. Deciding how to alleviate the situation can be tough, and you may be faced with having to get rid of the critter in some way, or if the troublemaker is a male, perhaps castrating him. That may be a very difficult decision — what if that male is your best stud and has produced perfect offspring?

What many forget when it comes to animals is the need for a thick skin in order to deal with the harsh realities of life, and especially death, on the farm. Some newborn animals just don't survive. Sometimes you have too many babies in one year — more than your farm can support — and you have to give some up. Or favorite animals can become sick or injured and need to be put down. If you choose to have animals on your farm, you can expect to incur losses every year. Are you ready for this harsh reality?

Using products from animals

It can be particularly cool to be able to say that the sweater a friend is admiring is one that you hand-knit from yarn that you hand-spun and hand-dyed after gathering the fiber from the goat you raised. Chapter 17 explains how to get fiber from the back of an animal into something you can use to make beautiful garments or accessories.

If you're going to expect meat from your animals, you need to find a reputable butcher or dress out the animal yourself. But in either case, you know where the meat came from, what the animal was fed, and how he or she was treated. See Chapter 16 for help in finding a butcher.

Good care can go a long way to a better-tasting meat, milk, or cheese. As the commercial says, "Great cheese comes from happy cows."

Planting a garden and raising crops

In addition to enjoying fresh country air, reaping the fresh fruits and vegetables of your labor is part of the fun of embarking on a hobby farm in the first place. Growing your own food can be so rewarding. You can't beat garden-fresh fruits and veggies, and preserving produce allows you to enjoy the harvest year-round. (See Chapter 15 for some ideas on using and preserving food items.)

Although some crops do well only in certain climates, you still have an abundance of choices for which fruits and veggies to grow on your farm. Your biggest limitation is how much work you want to put into the venture. For the best results, check with your local cooperative extension service (www.csrees.usda.gov/Extension) to see what grows well in your area.

Look for seeds at the local country store or go online to find an even wider selection. You can get your plants started on a healthy life in a lot of ways, from starting them in a greenhouse to using compost and fertilizers to dealing with pests. Chapter 13 helps you decide what to grow, and Chapter 14 helps you get those plants started and names ways of dealing with pests that threaten your crops.

Going organic can be a little more work, but you may discover it's healthier for you and the environment, and other people may be willing to pay extra for your tomatoes if they're certified organic. *Organic* means using only natural materials in your crops. In other words, the fertilizers and pesticides you use contain ingredients that originated from *organisms,* or living things; no artificial chemical materials are allowed. See Chapter 14 for some help with going organic.

Having some fun: Agritourism and the social scene

Part of what you choose to do with your farm may be related to how you want to interact with people and expose them to farm life. Get people to come to your farm and show them what life is like with a farm tour, or let people come and pick their own fruits or veggies.

Country folk enjoy some home-grown activities that just don't happen in the city, such as the county fair or farmer's market or a good old-fashioned barn dance. You can certainly run a vegetable booth at the market or set up a dance in your own barn. Farms offer so many opportunities for having fun that don't cost much at all. See Chapter 19 for some ideas.

Considering a location for your farm

How much acreage can you handle? The more, the better may sound great at first, but consider that you have to maintain that land, which may mean doing a lot of mowing, controlling weeds, planting, harvesting, and even fence building. On the other hand, if you plan to start your operation small, you may want plenty of room for expansion.

Is there anything exotic (either animal or vegetable) that requires a specific climate? What you want to grow or keep on your farm can impact where you locate. For example,

- ✔ If you absolutely love Walla Walla onions, you need to live in the northern part of the country; if Vidalias tickle your palate, the south is for you.

- ✔ If you think *qiviut* (the fiber from a musk ox) is the ultimate, you need to be in the colder northern climes.

- ✔ If you want to grow exotic flowers, you may consider an area that's warm year-round (or invest in a greenhouse).

The possibilities are endless. Even if you decide to stay in your current state, not venturing too far away from family and friends, you have to decide which part of the state to settle in. Different areas have different natural features, soil quality, and the like.

Even within specific areas, different zoning laws may apply. Will you even be able to have animals on your land? Can you get a permit to build a barn as big as you want it to be? Chapter 3 talks about permits and zoning laws.

Finding the right amount of land in the right location with a perfectly acceptable house already on it is certainly possible; however, if your needs are pretty specific, you may have to build from scratch, which often means building your own dirt roads, getting a propane tank, and having someone dig you a well and septic system. Chapter 3 discusses some of what you have to consider when picking out a location, and Chapter 5 lists some of the outbuildings you're likely to need to build.

Preparing to Take the Rural Plunge

So maybe you've examined all the pros and cons of hobby farming and have decided it's exactly what you want at this time in your life. Good for you! So many have gone before you (including me!), and it's a wonderful lifestyle if you're up to it. In this section, I discuss some ways to get yourself ready before you start hunting down properties.

Getting your feet wet

Dreaming about the joys of farm life is often a lot different from the reality of managing an actual farm. However, you can get a feel for the lifestyle before you actually dive in. Try some of the following ways to get your feet wet:

- ✔ **Rent a farmhouse for a year or two.** Spending time in the community is a great way to figure out whether living way out there works for you.

- ✔ **Take a farm vacation.** Stay in a guest house on a farm where the fun activities include actually doing some of the farm chores.

- ✔ **Chat with a farmer.** Talk to a local farmer who's out in a field as you drive by and ask what he or she likes and doesn't like about the lifestyle. Strike up a conversation at a farmer's market. Attend a fair or a seminar about agricultural issues, and chat with the people around you.

- ✔ **Browse in a country store.** Visit a store that sells animal food and gardening supplies to farmers. See what kinds of tools and accessories you'd need if you were to take the plunge (I discuss tools in Chapter 6). Talk to the salespeople or even to the customers, and check out the community bulletin board by the doors.

- ✔ **Just go for a drive in the country.** Go out on a Saturday afternoon and observe what's going on.

- ✔ **Peruse books, magazines, or Web sites that cater to farmers.** Get an idea of the kinds of tasks farmers deal with on a daily basis. I list some resources in Appendix A.

- ✔ **Take a class or workshop.** Several places offer courses designed to help the prospective farmer get some skills with plants, animals, and machinery. They explain what to look for in soil, how to deal with financing, and even provide help with marketing. For instance, you can take online courses specifically for beginning farmers from Cornell University (`beginningfarmers.cce.cornell.edu/onlinecourse.html`) or Penn State (`bedford.extension.psu.edu/agriculture/BeginFarmer/FarmCourse.htm`).

  Or attend free or inexpensive workshops or seminars on specific activities, such as composting, sheep shearing, milking, or organic farming. Check the local papers or do an Internet search to find opportunities in your area.

- ✔ **Get a mentor.** Look into programs such as Iowa's Farm On program, which matches beginning farmers with old pros who plan to retire (`www.extension.iastate.edu/bfc/programs.html`).

Do plenty of research, but don't feel you have to be an expert in everything farm-related. Use your resources — read, call your cooperative extension office, talk to farmers — and then prepare yourself for a little trial and error and dive in!

Looking up local laws

Before you start anything, you have to know what's copasetic, kosher, cool, or just okay with the local laws. What you want to do isn't necessarily what the laws let you do. Check with the city or county before you embark on your endeavors. See Chapter 3 for info on dealing with zoning laws and the permit process.

Something that may seem innocuous to you may be illegal because it's really, really bad for the environment. For instance, capturing and keeping deer can lead to fines or worse. Get in the habit of checking laws.

Chapter 2

Before Taking the Plunge: Is This the Life for You?

Most people who decide to live in the city do so because they think the benefits of city living far outweigh country living. For one thing, you can find cultural activities around every corner, which is something you have to say goodbye to if you go for a bucolic lifestyle. A city also has a life beat, a pulse that some people can't live without. Living in the city can be fun, with lots of activity and life and diversity. However, for those who don't like the rat race, traffic, and smog that come with living in or near a big city, country life can seem like a dream come true.

But some trade-offs may prove to be too much of a hurdle. Consider all the ideas I discuss in this chapter (and the other chapters of this book) before you take the plunge. Getting away from the city is probably what you want — after all, you're reading this book! — but maybe going all the way into the country life isn't *really* what you're looking for. This chapter helps you sort out the benefits and drawbacks before you start putting all your eggs in one basket.

Knowing What Draws You to the Country

Think about your reason for the change. Are you just trying to get away from traffic and smog, or are you really interested in becoming a farmer? Your honest answer can make the difference between jumping into a dream and jumping into a nightmare. This section helps you sort out your motivations for moving to the country so you can figure out whether that farmstead you have your eye on is likely to meet your expectations.

Returning to rural roots

Maybe your parents or grandparents lived on a farm, and now that you've made your fortune in the city, you think it's time to go back to your roots — back to a kinder, simpler time when making raspberry jam or homemade root beer was the entertainment for a Saturday afternoon and everybody had the best time. "Fresh air" meant cow manure and hay, and alarm clocks came in the form of a crowing rooster. Or maybe you're simply interested in a more peaceful way to live — a slowing down in anticipation of retirement.

Nostalgia can be a great draw, and country life still offers a slower pace than its city counterpart. Still, keep in mind that you may not be able to reproduce a childhood memory in exactly the same way. The cars and other motorized vehicles will be different, your favorite childhood toy may no longer be around, and of course, the penny candy is long gone.

Going green (acres): Independent or sustainable lifestyles

Fending for yourself after a lifetime of relying on others for things such as power, water, and even food can be very rewarding. You just feel good when you've accomplished something that provides you with something you need, pleased that you did it all by yourself. Becoming completely self-reliant is often impractical, but on the farm, you can find new ways to increase your independence every year.

Or going back to the farm and making do for yourself may be your way of being kinder to the environment. For every tomato you grow yourself, that's one less that you'd be buying from a store. (A tomato sold in a store gets packaged and transported across the country.) You also know what kinds of chemicals, if any, were put into the ground to grow that tomato.

Trying something new

Maybe you're interested in finally being able to have that horse you always wanted as a kid. You can't have large animals in the city, and even in most suburbs, zoning laws and space at the very least restrict the number of animals you can have. And some animals — such as pigs or even chickens — you can't have at all with very close neighbors.

Or maybe you've always dreamed of having a lot of land. The closer you are to the city, the more expensive land is. You can get a whole lot more land for your money if you go farther out. If you're truly interested in farming, the best place for that is, well, a farm.

Trying farming may turn out to be the best thing you ever did, or it may not turn out as great as you dreamed. To test the waters, consider renting a farmhouse for a year. That way you can experience the life without as big of a commitment. Or perhaps take a farm vacation (kind of like a ranch vacation), in which you stay in a guest room and actually do some of the work on the farm during the day. Places across the country, such as the Pennsylvania Farm Vacation Association (www.pafarmstay.com), offer this kind of experience.

Making some money

As a hobby farmer, the richest thing you're likely to get your hands on may be the soil. Even in full-scale operations, turning a profit as a farmer is tough, and it requires business planning, hard work, perseverance, and a fair bit of luck. That's why hobby farming treats farming as, well, a hobby. Most hobby farmers have supplemental income from another job.

If a financial gain is your motive, even some of the animals you consider having on the farm may be a lot more trouble than they're worth in the end. Cows eat a lot — while they're growing, they eat their weight every month — and you typically don't sell a cow for meat until after he's at least a year old. That's a lot of food. And then you need the trucks for transporting the cattle (the cost to buy and maintain the truck is not trivial). Raising just a small number of beef cattle may not turn out to be as good of a money maker as you thought — it may even *cost* you money to do this.

Or consider fiber animals. For the amount of money and time I put into my animals, it'd cost a whole lot less to just buy the fiber from somebody else and forego the task of raising the animals. But the end result (the fiber) isn't the only reason I have the animals. I just like having them out in the yard.

A slower pace: Entering a new kind of rat race

The rat race of city life is dealing with rush-hour traffic, stressful jobs, and demanding bosses. Life on the farm still has its stresses, but they're very different. On the farm, you have to deal with vermin (rats and other pests) in your barn, getting into the feed you've set aside for your animals and into your house as well. (We had a surge of mice in the house when construction started on a nearby subdivided lot. With a loss of outdoor mouse habitat, they flocked to our house. Thank goodness for our cats!) Staying on top of the rat, mouse, and gopher problems can be a challenge.

Some things are more peaceful and quiet (less traffic, no neighbors 10 feet away), but country life isn't entirely stress free. Besides the literal rats, you encounter other stresses, such as power outages, longer drives to get to necessities such as groceries and gas or medical attention, and running out of propane in the middle of winter. There'll also be those early morning forays into the animal yard to feed your critters before you get to eat and sometimes another trip in the evening before dinner. You have to stay on top of cleaning out stalls and barns, keeping fences intact, and maintaining farm equipment. And don't forget those winter storms that mean you have to plow or shovel pathways so you can get out to the animals.

In other words, in moving to a farm, you're still in the race, but you exchange one type of rat for another.

Raising responsible kids

Maybe your grandparents lived on a farm, and you remember the fun of visiting when you were a kid. There was so much open space, and riding tractors and helping with the animals were so much fun! Chores didn't seem like work because cool machines and animals were involved, things way different from what your regular daily routine was like in the city. Now you feel that would've been a good way to grow up, so you want to move into the country to give your kids that lifestyle. Being out in the country really can have some great rewards for the kids:

- ✔ They learn about hard work and responsibility maybe a little more easily than if they were to live in a city apartment, where fun may be hours in front of a video game.

- ✔ Less traffic means that playing near the streets is safer. Or the back pasture is even better than the street for things such as figuring out how to ride a tractor or four-wheeler. (But you still need to instill in your kids a respect for vehicles on the road — some people like to go really fast on country roads because there are fewer intersections and stop signs.)

- ✔ Being outside is great for kids of any age, and letting them be outside all day in the country often feels safer than in the city.

- ✔ Kids can raise animals, ride horses, and shear sheep and llamas. Children are likely to get involved in 4-H projects, giving them pride in what they've accomplished with those animals.

- ✔ They can discover how to grow and harvest plants, taking pride in growing plants just like they do with the animals. Perhaps their special attention to the pumpkin crop will lead to a prize-winning pumpkin at the state fair.

Farming is good old-fashioned fun that can prove to be a bigger reward when you have to work hard to get at it. Still, simply because neighbors are farther away, your kids probably won't have the opportunity to round up some friends and start up a basketball game at a nearby park or gather at the city swimming pool. But the good news is that they're likely to do things that are more creative than hanging out at the mall or spending hours playing video games.

Don't get a false sense of security that your kids will grow up wholesome and trouble free simply because they're exposed to hard work and responsibility. Studies show that alcohol and drug use tends to be as high among rural teens as it is among their city counterparts, so you still need to parent them well. As you would anywhere else, take an interest in their lives, and pay attention to what they do, who they hang out with, and where they go.

Considering Practical Issues before Getting a Hobby Farm

Getting away from the city can be great, but you have to consider a few aspects of country living before you jump into hobby farming. You encounter some very good, enjoyable elements — wide open spaces, the joy of eating vegetables that you grew yourself, bringing new breeds of critters into your family — but some of the consequences of moving away from the city can be a little harder to deal with: power outages, longer drives to the grocery store or to satisfy a midnight Big Mac attack, or maybe having to use alternative power sources.

Living in the city and being able to rely on somebody else to supply things such as power or all your food makes life easier in some ways. But when you're on the farm and making do for yourself, some tasks you used to take for granted can seem like a major chore. In this section, I explain some of the responsibilities and practical issues you can expect on a farm.

Early to bed, early to rise: Doing a day's work

Taking on the responsibilities of a farm means extra work, and if you also have a regular day job, that's like taking on a second job. You need to do some farm tasks, such as milking or feedings, in the morning, so you have to get up especially early, which usually means getting to bed early so you get enough sleep.

But even if you don't have jobs that need to be done at the crack of dawn, you still have to put in more hours than you likely would without the farm responsibilities. You may have to feed and water your animals both before you go to the office and when you return home. Other regular chores include collecting eggs, cleaning stalls, and dealing with iced up water buckets on winter mornings.

You can choose to do everything and keep working until the day is done, or you can pace yourself and just do the tasks that need to be done at the moment. You can choose *not* to work that hard if you want to — that's one of the joys of farming as a hobby. But some tasks are more urgent than others:

- ✔ **Planting and harvest:** Plants have to go into the ground within a certain timeframe — during planting season. If you miss that window, you don't get a crop that year. When the plants are ready for picking, you have to do it in a certain timeframe, or your crops rot on the vine.

- ✔ **Birthing:** In birthing season, bringing the little ones into the world can take every spare moment. Not all births go smoothly, and there's always some emergency that you need to take care of at any hour of the day. Some animals (horses in particular) like to have their babies in the middle of the night, and that means you may not get much sleep during foaling season.

- ✔ **Machinery upkeep:** If machinery essential to your daily operations breaks down, you have to fix it right away to keep the farm running smoothly. And machinery *does* break down, often at the most inconvenient time. You have to get right to the task of fixing harvesting equipment or the backhoe (if you're puttering around in the back pasture building a road) when it does break down. Or if the snow removal equipment breaks down after a huge snowfall, you can't get around on your property if you don't fix it. You can't put off these things until another day.

Getting a handle on many of the skills you need to use on the farm — such as raising plants, caring for animals, or running farm equipment — is helpful. But it's not necessary to be an expert, so don't hold off on moving to the farm simply because you don't know everything. A lot of the skills you need can be learned on the job, with some support from books, neighbors, or people at the cooperative extension service. When you first start your operations, keep them small enough to manage and be prepared for some trial and error.

Getting out and about

The wide open spaces of the country and the number of chores waiting at home can make leaving the farm a bit of a hassle sometimes. Here's how getting out and about can differ from when you live in the city or suburbs.

Taking trips to town

When you live out in the country, going into town can become a chore. The trip is often a few more miles than the one you used to have, and you may not always want to go, even to take part in fun things. Also, you and your kids are likely to miss out on some of the good things the city has to offer, such as more cultural activities — museum visits or exposure to the theater. Of course, you can still do those things, but they'll require a longer drive.

Get around the hassle of the long drive by combining chores. I never go into town for just one errand; I gather them and then go when I have three or four things to do. And of course, you may want to double up with neighbors and carpool to special events.

Planning a vacation

If your farm requires daily attention — for instance, if you have animals who need to be fed and watered every day — going on an extended leave may be a bit harder. That two-month journey across Europe that you've always dreamed of may not be possible. If you _do_ decide to go, you have to hire somebody to come over to the farm every day to feed the animals, water the plants, make sure the automatic watering systems are working, or whatever else you have going on that needs regular attention.

A good place to look for a farm-sitter is just across the fence. Perhaps a responsible teenager lives nearby, one who's already versed in the ins and outs of farm life and would love a little extra money. Or check with the local high school, ask a local vet, or put an ad at the local feed supply store.

Staying healthy

The countryside may sound like a recipe for good health with lots of fresh air and room to run, but health is still an important issue. Here are some medical considerations.

Minor emergencies can be major ones

Because you're farther away from help, what may otherwise be a minor emergency can turn into a major, life-threatening one. A fall from a tractor can result in lacerations and broken bones that, if you tend to them right away, result in a time of convalescence. They're an inconvenience in that you have to slow down for several weeks while you heal. But if you're all alone in the field when the accident happens, with nobody else around to notice, and you're unable to get yourself to safety, you can end up losing a limb or dying from blood loss or shock.

To ensure you can get help in case you have an accident, work in pairs or at least carry a cellphone.

Allergies can ruin everything

Your whole plan can come crashing down if you have allergies you can't deal with. You may be allergic to animals or to the hay you feed them, in which case you have to find somebody else to do the feeding. Or you may have an allergy to whatever plants are growing in the area. On a trip to Austin, Texas, during January (when the cedar trees were dispersing their pollen into the air), I discovered that I'm very allergic to cedar pollen. Had I found a dream home in an area where that type of tree was prevalent, I would've been moving into a nightmare!

Visit a working farm to see whether you have any allergies you don't know about (animals, hay, and so on). Or after you decide on an area that attracts you, visit it in all seasons before making the final decision. After all, cedar problems happen in January, but hay fever is usually in the spring or fall.

Looking at differences in gas and electricity

Powering up your appliances and farm equipment may not be as easy as plugging something into an outlet. Here's how your electrical and heating utilities may differ:

✔ **Generating all your own power:** If you're in an area that has no utilities or services, you have to come up with alternatives such as propane, a windmill, or solar power for your electricity and heat. (See Chapter 7 for some ideas on working with alternative power sources.)

✔ **Paying for new power lines:** Even if you're close enough to on-the-grid power suppliers that you can use their services, perhaps no current connection is available. In that case, you have to go to the expense of having lines dug out to your place so you can use those services. That can cost a couple thousand dollars!

✔ **Providing power for outbuildings and other remote locations around the farm:** You may have to do a little work here and there, such as put in a windmill to generate electricity for a corner of the farm that's far from your house.

✔ **Preparing for power outages:** If you do get your electricity from the local power company, be aware that power outages do happen (they happen at my house a lot more frequently than I'd like). You have to either be able to deal with being without power for a while or install some sort of generator to hold you over until the power comes back on.

Be prepared to take care of your own power-generating equipment. You may not always be able to call somebody to come fix things. If you're using the windmill for electricity and it goes down, it's up to you to get it back in working order. You also have to monitor propane tanks so they don't run out (but in this case, you can call somebody to come out and fill it up).

You may also need your own well, propane tank, or septic system. For more info on how you may have to handle your own utilities when city services come up short, see the upcoming section titled "Things You Probably Don't Have Now but Need in the Country."

Dealing with bad weather

When you're in the city and bad weather comes along, the city is responsible for plowing the roads or cleaning up the streets after a bad wind or flood. Out in the country, you're likely to be responsible for the cleanup yourself. And depending on your situation, you may have to deal with a lot of snow, flooding, or a wind that knocks down your windmill.

Other weather conditions can threaten your buildings, crops, or animals. Droughts and hailstorms can cause havoc. When you depend on things that depend on the weather (you grow crops for food that need water and sun, or you depend on a stream for watering your animals) and the weather doesn't cooperate, your financial situation — and your stress levels — can get serious.

Being at the mercy of the weather is a fact of life on the farm with no real solutions. You just have to go with the flow (or lack of a flow in the case of droughts).

Being smart about the money

I'm guessing that just like me, the typical hobby farmer isn't into it for the monetary gain. It's more likely the joy of working with your hands and being out in the fresh air that drives you. So putting out a little extra money (and thus making your efforts not a money-making endeavor) is no different from putting money into any other hobby. It just makes you happy to do it!

But you still have to be smart with your money and not plunk down thousands of dollars for something you can easily, and much less expensively, have someone else do for you. In this section, I talk a bit about finances (see Chapter 3 for more information on managing your farm budget).

Having some financial nerve

You need to put out cash — and sometimes a lot of it — to keep a hobby farm running smoothly. Maybe you need to separate animals, so you have to buy and install fencing and/or gates. Or maybe you decide a backhoe or tractor is a toy you can't live without. Even if you plan to make a little of that money back by selling veggies or animals, you still have a big outlay of cash in the beginning and then a long wait until the fruits of your labor pay off.

Because income can be delayed or even uncertain, diving into this hobby is not for the financially faint of heart. Be aware of what you can afford and how comfortable you are with any debt.

Divvying up the work: Using a farm contractor or going it alone

When you know the essentials of farm ownership, you still have to deal with all that equipment you need for running the operations you want to delve into. You may find that hiring somebody else to do some of these things may actually be better than attempting it yourself.

Running a farm can be expensive, especially if you feel you have to own, operate, fix, and maintain every piece of farm equipment needed to do every task. If you have only a few acres, buying some of the bigger equipment, such as a combine, may not be financially feasible. Hiring a contractor to come during harvest season and cut for you may be much more cost effective.

Or consider the greenhouse. Using somebody else's greenhouse for starting plants may be preferable to building and outfitting your own greenhouse. You can buy an awful lot of starter tomatoes for what it'd cost to erect and maintain a greenhouse for a year.

If finding somebody else to do some tasks for you ends being the best option financially, here are some resources for finding contractors:

- ✔ Your neighbors
- ✔ The local feed supply store
- ✔ Your meat cutter
- ✔ The want ads

If you do decide to do some tasks yourself, you may be able to get by with some minimal equipment. We have a backhoe that has so many uses, and a friend uses a small tractor with a pay loader and mower bed that she says is sufficient for her needs; it mows the lawn and grazing fields, cleans out paddock areas, deals with manure, and can even dig a grave.

A few agencies offer low-cost lending options to help you finance that equipment you just can't live without:

✔ Farm Credit Services (www.farmcredit.com)

✔ Farm Service Agency (www.fsa.usda.gov)

✔ Small Business Administration (www.sba.gov)

Things You Probably Don't Have Now but Need in the Country

Sometimes the area you move into doesn't have all the services you need to live comfortably. You may not be close enough to power lines or the sewer or water pipes to be able to use those systems. The good news is that you don't have to pay for the services. The bad news is that you have to provide your own services — and repair them when something goes awry.

You may have to have someone build (or erect or dig) wells, septic systems, or alternative power supplies. Or you may responsible for repairing weather damage (such as washed-out roads) or hauling your garbage to a landfill instead of having it conveniently picked up at your curbside. This section runs through some of the adjustments you may have to make.

A well

Many if not most farms have their own wells. The land in remote rural areas simply isn't close enough to make it financially feasible to connect into the city's water supply.

With a well, you don't have a monthly bill to the city, and barring a severe drought, you aren't restricted to a certain amount of water use — and you don't get dinged for a fee if you go over that limit. The bad news is that you may have to dig that well, which can be expensive if you have to dig several hundred feet down to get to the water. You need to hire someone with the proper know-how and equipment to do that. Even after the well has been dug, the water quality may not be up to snuff for human consumption. In that case, you have to either drill a new well in another location or install some sort of filtration system.

You also have to maintain that well. We had to replace our electric *well pump* — a mechanism that lives in the well itself and suctions the water up and into the pipes — to the tune of several hundred dollars.

You should periodically check the well water to be sure no contaminants have crept in (the EPA suggests an annual check). You can find someone to do it (check with your county offices) or get a kit to do it yourself.

You also have to face the fact that your well *could* dry up. If that happens, you may be without water until you make some other arrangement. For us, there's been enough subdividing and building in the area that, if we want to or need to, we can hook into the city's water system.

A septic system

If you can't hook into the city's sewer system, you need your own means of disposing of human waste, and that means having your own septic system. You need the experts to hook up everything so it's up to code. For a new property, you may dig your own and have a plumber hook everything up.

How it works

In a septic system, waste goes out through your pipes and ends up in a holding tank, where good bacteria break it down. As the tank fills, it overflows through a series of level pipes, where it's filtered down through the soil and distributed to a large underground gravel bed that's at least a foot deep. (It's well underground, so you never see or smell it.) Figure 2-1 shows how the system works.

Figure 2-1:
Wastewater goes into the septic tank and then leaches into the soil through the drainfield.

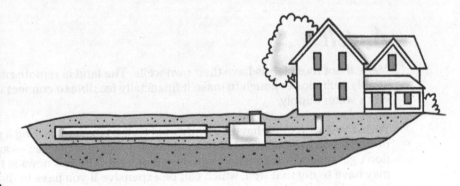

Building and maintenance

Before you even start digging a septic system, you have to get a *percolation test* done to ensure the soil is drainable and can support a septic system. For instance, if the ground has too much clay or rock in it, you have to move your proposed septic system to another area on your property or perhaps blast the rock and add a load of sand. Typically, you can find a place where the test passes. Without a means of dealing with household plumbing waste, you can't get a permit to build.

When planning on building a septic system, hire a planner to test several areas of the property to make sure a septic system can be built in more than one area. The septic system will have to be replaced in the distant future, and the new one can't be rebuilt on the same site.

Even if the property already has a septic system, you need to maintain the system. It's recommended that you have a serviceperson come out every year or so to pump it out (now there's a job I'm glad I don't have!). You can also purchase additives to pour down your drains and keep the septic system healthy.

The waste comes not just from the toilet but also from other plumbing pipes. You should really restrict what you put down the garbage disposal; our plumber told us not to put things like onion skins, grease and cooking oils, or big pieces of potatoes down there. Ask your plumber for suggestions concerning the washing machine and showers — anything that goes into the plumbing pipes.

A propane tank (or other alternative fuel system)

When you're too far off the grid to be able to hook into the lines fed by the power and natural gas companies, you have to come up with your own means of heating (and powering) your house and outside operations. Propane is probably the most common alternative power source, but you can choose from several other power sources as well. See Chapter 7 for ideas.

A generator

Power outages are a reality, whether you're using the local power plant or relying on your own wind or solar systems. Usually, you have to get the power back on quickly. Too long without heat in the winter can freeze pipes and people; no electricity in the summer can mean your automatic waterers don't work and food spoils. If you're on a well and using an electric pump, no electricity means no water — and no water means no toilet flushes!

A generator can alleviate those problems by keeping at least minimum electrical operations online. But running with a generator is more expensive than other means of providing power, so use it only as a temporary, emergency power source.

Generators are typically diesel engines that convert mechanical energy into electricity, but you can also get a special contraption that doesn't have an engine itself and that, when hooked up to your tractor, does the same job.

A new way of taking out the trash

The days of wheeling your garbage can to the curb and having the city come and haul it away may be over. Depending on your location and what the city is prepared and willing to do with your trash, you may have to haul it to the landfill yourself.

Many rural communities don't offer curbside pickup. A friend uses a burn barrel (or sends bags of garbage home with her suburban relatives!), but open burning can be dangerous and is environmental taboo, so be careful and check with local laws before burning your trash. We're fortunate to have close garbage pickup. The garbage truck won't come down our dirt lane to pick up at the house, so garbage collection isn't exactly curbside (well, we don't have curbs!). We have to keep our garbage cans about ¼ mile down the road at an intersection where the garbage truck *will* come.

Firearms

Whether or not you like the idea of guns, they do have a place on the farm. They're very handy for protection against predators, both the kind that threaten the humans and those that threaten the animals on your property. Although laws protect wildlife, sometimes you don't have a choice. If a coyote gets into the pen with the newborn lambs, the only way to save those lambs may be to sacrifice the coyote.

Besides dangerous wildlife, you may get visits from packs of wild dogs. Sad but true, some people abandon unwanted puppies and dogs out in the country. These dogs then turn feral and have to fend for themselves, often hunting and eating whatever they find. It's all about survival. Your small animals look like easy eatin's to them. You may feel sorry for their plight, but you need to do something about them so your flock (and the hard work and expense you put into raising the animals) doesn't disappear without your getting any benefits. Intervention is probably the only way for you to get peace (and peace of mind).

See Chapters 4, 9, and 14 for more ideas on dealing with wildlife and other pests and for a discussion on firearms on the farm.

Chapter 3

Finding Your Place in the Sun

- -

In This Chapter

▶ Considering how much you want to spend

▶ Settling on a location

▶ Figuring out how much land you want

▶ Evaluating the property

▶ Buying or building

- -

You can settle in many places in the world, depending on your interests. Some animals or crops don't do well in certain climates and thrive in others, and depending on what you want to do, you may have to make a big move. So knowing that, you need to find a place that accommodates your interests while keeping you close enough to civilization (and keeping the commute to the job that helps pay for this hobby sufficiently short).

Obviously, jumping into a farming venture, even on a hobby-sized scale, means more than just finding some land and putting a house on it. In this chapter, I discuss what to think about when looking for your perfect property.

Establishing and Sticking to a Budget

You pour money into hobbies because you love doing them, and most hobby-ists are lucky to break even. Decidedly, a hobby farm operation isn't your main source of income. Hobbies, by definition, just don't do that. If this were a for-profit venture — something you go into not only to feed your family but also to prosper — your decisions would be quite different.

A hobby farm does involve things you may spend a lot more for than you would if it were a verifiable business. Case in point: We bought a used back-hoe, and it cost a boatload of money. Based on the income we get from the farm, that expenditure wasn't even close to being justifiable, even in horse-shoes, were we counting on farm activities being good to our bottom line. But hubby wanted one, and it was — and continues to be — one of the most fun toys he has ever acquired.

However, you have to be realistic when you're planning what to do on your farm and what's affordable. If you're looking at property that doesn't have a barn, should you plan to spend several thousands of dollars to erect a barn so your animals have a place to go in bad weather? (Of course! They're your babies!) Should you put out the bucks for the backhoe so you can do your own landscape work? Should you buy expensive harvesting equipment so you can easily and quickly (without a lot of *your* sweat equity) pick all the fruits and vegetables you're going to so lovingly coax to fruition?

Before you start looking at properties, sit down with your significant other to discuss your budget. Decide what you want to do and make choices concerning what you'll put the money out for. You'll probably want to start small and grow as your skills and your commitment to the lifestyle grows. Keep in mind, too, that the initial outlay isn't the only thing you have to cover. There'll be operating costs, recovering from mistakes, and repairs. And if you're planning to sell some farm products, there'll be storage needs, packaging, transport, and marketing.

Deciding Where in the World to Settle

Think about what you want to do with this incredibly consuming (in terms of time, money, and emotions) and incredibly rewarding hobby. Do you want to grow oranges? Do you want to raise musk oxen or trout? Whatever you decide profoundly influences where in the world you should locate your hobby farm.

Here are some basic elements to consider:

✔ **Climate:** Some plants and animals do well in certain regions and don't do well at all in others. Oranges don't work in Alaska. Musk oxen don't work in Florida. Tomatoes grow great in most of the Midwest but do very poorly in the South or Northwest. Llamas are popular hobby farm animals in Michigan and Western mountain states, but they don't do well in Alabama. Climate is a big deal and is a big factor in your decision, so make sure whatever you want to do is doable in your chosen region.

✔ **Environmental preference:** Consider your personal inclinations. Where do you feel most comfortable, or where do you envision your dream location? What type of environment do you enjoy living in and near? Do you like mountains, oceans, plains, grassland, or rolling countryside?

✔ **Distance from family:** Do you want to be close to relatives? This is a number one consideration for a lot of people. Regardless of where you see your dream property, it's not going to work if you feel it takes you too far away from family.

✓ **Level of isolation:** How isolated — or not isolated — can you handle being? In some people's dreams, being in the middle of nowhere is perfect. Of course, "middle of nowhere" is a subjective term. It may mean 100 miles (or 15 miles) from a city center, and whatever your ideal is can limit your choices a bit. Keep in mind, too, that the closer you are to the city center, the more you run the risk of having urban sprawl swallow you up.

After you decide on a region (maybe you really want to grow citrus fruits and you love the idea of no snow, so Florida is your dream place), visit several times. You can see what it's like to actually live there in the hottest part of the summer or find out whether you're allergic to something in the air there.

Amount of Acreage: Deciding How Wide You Want Your Spaces to Be

How much land do you really need or want? Certain operations require a lot of land; others, not so much. In particular, crops and animals, and their respective sizes, require different amounts of space. And the more land you acquire, the more you have to maintain. I let you know what to look for in the following sections.

For growing various crops

How much land you need for crops obviously depends on how much you want to grow. Consider how much you want to plant and how many people you intend to feed. Just your family? Family, friends, and co-workers? Or do you want to have enough left over to take to a farmer's market and possibly make a little money on the side? You can get by with a really tiny corner of a yard if you just want to grow enough to have a few occasional fresh vegetables for your own table.

Here are the basics on what you can reasonably plant, given your land size:

✓ Greens needing the smallest space are herbs. You can even grow them in pots.

✓ With a couple of square feet you can grow salads. Plant some lettuce, tomatoes, radishes, cucumbers, carrots, and so on.

✓ With a 10-foot square area, you can expand into beans, leeks, and turnips.

✓ If you have at least 500 square feet dedicated to your plants, you can grow about 90 percent of a person's annual veggie requirements!

To maximize your plot, consider not growing fruits and veggies that take up a lot of space and are easily accessible at the store, such as potatoes.

Your harvest will vary depending on the weather, your land, and your farming practices, but if you want to plant multiple acres, try to find out your neighbors' crop yields in the recent past or look for county farm records. You can go to www.nass.usda.gov/Charts_and_Maps/Crops_County to find maps of the average county yields for corn, soybeans, wheat, and other crops.

As you evaluate how much space you need, remember that most veggies and fruits need a lot of sun. That's why the majority of your garden space should be shade-free. Plants also need to be in a place where large trees won't be hogging all the water.

For keeping animals

The amount of space you need for your animals largely depends on local zoning laws. Some places let you have lots of animals, so the number you have is restricted by other things, such as availability of grazing forage. Some places allow animals on as little as ¼ acre.

But besides zoning laws, you have to consider what's around in the way of pasture. Three acres with no grass on it is not the same as 1 acre with good, healthy grass. Also, there's a limit to the amount of trampling that grass tolerates. If you have only a small space, the grass gets a workout and may possibly dry out, leaving nothing for the animals to graze on.

Getting a new lease on rural life

Instead of buying, you may consider leasing land from a nearby farmer and working it for yourself. Used to be a good chunk of the people who farmed land were just tenants, and somebody else actually owned the land. This practice is still going on today.

Individual leasing agreements provide info on the duration of the agreement, payment schedules, how you can use the land, and so on. For instance, the owner may stipulate that you leave a certain number of acres as permanent pasture, that you avoid certain crops, or that you take measures to control erosion to keep the land productive for the long term. In turn, the owner usually takes care of major repairs to the farm buildings, handles property taxes, and the like. Online classified sites, such as the one at www.agriseek.com, can point you toward some leasing opportunities.

Gardening for victory

Farmers have been growing their own food as well as growing enough for the whole country for many moons. Growing one's own food was typically left to those who had farmland, and city folk didn't bother with it. But during the World Wars, the U.S. government encouraged people to plant *victory gardens* in order to help with the war effort by reducing the pressure on the public food supply. The campaigns encouraged planting as a family or community project and as a new pastime. It was considered a national duty. People started tilling up little corners of their land, and by 1943, the victory gardens were responsible for 30 to 40 percent of the U.S.'s vegetables!

These days, a lot of people continue the tradition even though they may never have heard of a victory garden. Some cities even rent out planting beds so those who don't have a yard can still have gardens.

Although raising your own veggies may not seem like much of a national duty anymore, many people think they owe it to the Earth to be as close to their food sources as possible. Growing your own food can even help fight global warming! Really! Think about it: If more people were to grow their own food, there'd be fewer big trucks on the road hauling veggies to stores. That means fewer emissions of greenhouse gases. It also means less packaging that has to be made and then disposed of.

 A good rule of thumb is to give large animals like horses or cattle about 1 acre per animal. Llamas and miniature horses can go with a bit less, maybe two animals per ½ acre. Llamas don't like to be alone, so you should always have at least two. (See Chapters 10 and 12 for more about keeping llamas.) Sheep and goats need even less space than that.

You may even want to consider getting enough land to grow your own hay. Most large farms do that, but smaller farms tend to get hay from someone else.

Examining the Property and Its Location

So you found the (maybe?) perfect spot to settle and start up your new farming hobby. A stream runs through the land, there are mature trees, and you have several acres, more than enough to accommodate whatever you want to get into. But what about all those *little* things that add up to *big* nightmares? I cover the basics in the following sections.

Proximity to towns, schools, stores, and neighbors

Being far away from everything may at first sound like your own personal Xanadu (if it's been a while since high school, that's the Mongolian emperor's luxurious garden-filled estate in Samuel Taylor Coleridge's poem "Kubla Khan"). But when it comes right down to it, being in the middle of nowhere can be a huge problem instead of an idyllic dream world. Here are some concerns you may have about keeping your distance:

✔ **Schools:** How close is the nearest school? If you have kids, distance to the school is a big deal because if the buses don't come out your way, you have to provide their transportation to and from school (and soccer and flute lessons).

✔ **Towns and stores:** Being way out there sounds like a dream come true sometimes. But what happens when you're getting ready to entertain 20 people and you realize you forgot the hamburger buns and the nearest store is 45 minutes away? Yikes.

Or what about those times when you just want to see a movie or go bowling? You may have a really long drive to get there. Also consider whether you want to deal with the long drive home at night, maybe even in the snow.

✔ **Neighbors:** Being down on the farm, where a single house sits on dozens of acres, means your neighbors aren't very near. Perhaps that sounds good to you. But what about your kids? They may miss having other kids nearby for an impromptu game of basketball or hide and seek.

Or what about the feeling of isolation? I usually like my little slice of heaven out in the middle of nowhere. But sometimes, like when my husband is away and I'm out there all alone at night, I start to think I'd like to have a few neighbors a little closer.

Water sources

Water is life, and humans (and animals and plants for that matter) can't survive for very long without it. Deciding to live in the middle of nowhere often means providing your own water. After all, you likely won't be close enough to a city to get on the grid and hook into its water system. And keep in mind, too, that farm operations typically need a lot more water than just a house does.

If your property doesn't already have a well, you'll most likely need to dig one. The dig may be easy (good ground with a water table that's just far enough down that it doesn't stand the chance of going dry or being contaminated by runoff) or very difficult and expensive (very rocky ground with a water table that's a few hundred feet down).

On my farm, they didn't hit the water until they'd dug a little over 200 feet. And when the bill was calculated according to how far the digging went down, it added up to a lot more than we thought we were going to have to pay. But what are you gonna do? You have to have water. When looking for your perfect place, keep in mind there'll be an extra cost (up to several thousand dollars) if you need to dig a well.

Drainage

Although the idea that you shouldn't build in a flood plain or in a dry wash may seem obvious, not everyone adheres to the rule. A new housing development near me has gone up dangerously close to an old riverbed. All they need is one catastrophic heavy rain, and those houses will be ruined.

But even if you're not in a flood plain, you need to be sure the soil has adequate drainage so any heavy rains don't cause a pond to form in your yard — especially near the house or any outbuildings! And you have to consider your septic system if you're far enough out that you need your own means of household waste removal. The septic tank needs to be able to release the wastewater out into a leaching field so it can be absorbed into the soil underground. If the soil doesn't drain well, the leaching field won't be able to accept the wastewater properly and you'll get flooding or sinkholes. See Chapter 2 for more information on having a septic system built.

Roadways and access

You have to be able to get *to* your dream lot! You may find a beautiful, large, very inexpensive lot with a view, with a stream, with lots of mature trees, your oasis in the mountains. But what if the property lacks existing roads and you'd have to build a several-mile access route to the lot? You shouldn't forget the idea of the mountain homestead, but recognize that building the road is going to require permits and add to the cost.

Boundaries and fences

If you're going to have animals (or if you're going to have a small garden that you want to keep animals out of), you need some sort of boundary markers — either fences or natural barriers such as streams or rock outcroppings.

Fences can be expensive to put up if you have acres and acres that you need to protect. In days of old, ranchers just collected downed tree limbs (or cut their own) and erected crude but effective fences. These days, you'll likely have to buy materials.

Amenities for crops and animals

Besides the space for animals to run, you need a good source of water and fences to keep them in and predators out. And then you need barns or other structures to provide shelter for those animals and to store their food. You need enough space to accommodate those buildings as well as pasture space.

What kinds of animals you want to have comes into play when you're looking at land. Although horses like big spaces to run and exercise, llamas and goats love to climb and don't mind a hilly or rocky pasture — in fact, they love to play king of the hill. Other properties have lots of flat areas with few trees or other brush and are perfect for crops. The right features all depend on what you're looking for and what you want to do with the land.

Communication technology

Sometime in their lives, almost everyone considers moving to a place in the middle of nowhere. No traffic, no smog, no phones ringing. Just a peaceful life among nature. But the reality is that isolation isn't all it's cracked up to be. What if your house caught fire and it'd take the fire department an hour to get there? What if you had a heart attack or an accident in one of the animal's pens and had no phone to call an ambulance, which would take an hour to get to you, anyway?

Peace and quiet sound great much of the time, but as a rule, you really want to have some connection to civilization. Yes, in days of old, people may have managed without such help, but times are different, and you don't have to be all alone like that. In this section, I discuss connections to the outside world.

Phone service and cellphone coverage

Unless you're *really* out there, phone service is likely to be available at your house. You may even want to install an emergency phone in your barn. The phone service may not be completely reliable (we have outages periodically), but you can at least use the phone most of the time you need it.

Although not an absolute necessity, a lot of people rely on cellphones. When you're in your house, and if you already have a land line connected and working, you don't really need to have a working cellphone connection. However, a cellphone is a great backup to have with you out in the field in case some emergency happens and you've fallen and can't get up.

Internet access

Internet access can be a major concern if you decide on a supplemental career that you can do from home. The Internet has also become a kind of necessity for keeping in touch with friends and family. Okay, so people

150 years ago relied on the Pony Express and they were fine with it, but in this day and age when things don't *have* to be that slow, you may really miss out on a lot if you don't have the Internet. And what about some of the other great things the Internet gives you, such as weather or traffic reports, which are helpful if you're heading into town? You may also want to order books, advertise your farm, visit agricultural message boards, keep up-to-date on local, national, and international news, stay abreast of environmental issues, or engage in distance learning programs.

Keep in mind that Internet access comes in several different speeds and comfort levels, from dial-up and DSL to satellite, wireless, and cable. Satellite Internet is widely available, but it tends to be pricey. Dial-up is the slowest, and it can cause seriously bad words to fly if you're regularly trying to transmit and receive large files, but it does have the virtue of being available anywhere you have a telephone land line. As for your other options, every day it's getting easier to connect as new services become available. Check with the phone company or cable or satellite TV providers to find out how you can connect.

Even if you don't go with one of the in-the-house Internet access services, you can still get onto cyberspace by visiting the local library or even an in-town Internet café.

Local rules and regulations

Just because you want to do something on your land doesn't mean you're allowed to do it. See Chapter 4 for some discussions about fitting in and getting along with neighbors, and read on for info on local laws you may encounter.

Zoning

Even in the suburbs, you encounter zoning laws — rules about what you can build and what you can otherwise put on your land. In the suburbs, you run up against not being allowed to have horses or not being able to build a barn. In the country, you're more likely be restricted on things like

- ✔ How many houses you can erect (my little slice of heaven came with a rule of only one house per 5-acre lot)
- ✔ How tall a building or other structure can be — I've even heard of a restriction on flag pole height!
- ✔ How close to your borders you can put a barn

No matter where you settle, you'll have some sort of restrictions, but they're usually workable. Our barn had to be at least 20 feet from our neighbor's shared property line, but we had plenty of room, and moving it in a bit was no problem. Zoning laws differ not only by state but also by county and even in different sections of each county, and they change, so the only way to find out which laws apply to you is to ask.

Access and easements

Besides needing access to your own property, you may run into the issue of having to provide access to other people's property (called *easement*) if for some reason they have to cross over your property to get to their own. You can't erect anything (a structure, fence, and so on) that would obstruct any access route to somebody else's property, even if those would-be structures are on your own land.

An easement is about granting a right of way over your land. For instance, if a huge parcel of land is subdivided and a plot is bordered on all sides by other plots, each bought by different people, easements prevent the guy who bought the plot in the middle from getting shut in. For my farm, we're currently using our own well, and we have a septic system. But a lot of subdivisions have been coming in around us, and the city's lines for water and sewer have come closer. If we decide to be hooked up to those lines in the future, we'd likely have to go to the county offices and get an easement to dig across somebody else's property in order to reach them, or vice versa.

Sometimes, too, you may have to grant an easement to the city. In my neighborhood, easement is forcing some people to give up some of their property to have a road built. Before development began, the only roads here were narrow dirt roads. But with a lot of the land being sold and, with that, a lot of subdividing, we were encouraged to annex into the nearby city. However, after annexation, the city's rules suddenly applied. Roads had to be 60 feet wide and also had to have a curb and gutter, a grass berm, and a sidewalk. That meant giving up about 30 feet in front of our properties to accommodate these requirements. A few of the houses in the area were built too close to the boundary for that to be possible: Giving up 30 feet would've put the sidewalk in the front room of the house! Of course, there were negotiations, and nobody had to move their house back or have the sidewalk right outside the front door.

Property tax breaks with the Green Belt designation

Some areas can hook into a cool benefit called the *Green Belt Law* (or Green Acres). If you own a certain amount of acreage and use it for agricultural purposes, you can get a break on your property taxes.

Passed into law in 1974, the Green Belt program was meant to help farmers afford to keep their farmland by keeping property taxes down. Especially in areas very close to urban areas where encroaching development pushes up land prices (and thus property taxes), this benefit helps keep farmers in business and also slows urban sprawl by making it easier for farmers to stay and resist the temptation to sell to developers. Each state (and even each county!) has its own qualifications for achieving the status, but typically you need at least 5 contiguous acres that are actively used in agricultural activities, which may include raising plants as food or animals for food or fiber.

The tax breaks are significant, but the downside is that if you ever sell to somebody who doesn't use the land for agricultural purposes (such as a developer who subdivides the land), the new owner may have to pay rollback taxes. In Utah, for example, that means paying the difference between regular property taxes and Green Belt taxes for the past five years. That can really add up and may be a problem when it comes time to sell.

Some states use a similar program called the Conservation Reserve Program (CRP). With CRP, you agree to keep your land natural, with no farming or animal grazing for a certain number of years, in order to restore the environmentally sensitive land. The government pays you each year that you agree to do this. If the land is sold and developed before the agreed upon time, then the buyer has to pay back all the CRP money that was paid out over the years.

The grandfather clause

You may run into something called a *grandfather clause* when you look at what you're allowed to do on your land. It's a little clause stuck on the back of some laws that gives people some wiggle room when it comes to obeying a new law. This exemption clause means that even though the rules have changed, if you were doing something before those changes, you're allowed to continue your operation despite the new law.

The idea of the grandfather clause came about in the late 1800s and early 1900s when people were trying to deny the right to vote to certain classes of people following the Civil War. Lawmakers passed this clause to allow the vote to anyone who already "owned" that right as well as their descendants to exempt these people from paying poll taxes, taking reading tests, or having to otherwise meet new voting requirements.

Today's grandfather clause has nothing to do with voting or discrimination, but it still allows practices that were once perfectly within the boundaries of the law to continue even if the law has changed. It means no new operations using the old ways can be started, but operations that are already established don't have to shut down.

For instance, a law in Michigan, passed in 2004, prohibits keeping large carnivores (big cats and bears) as pets. But people who already had a pet bear or lion were allowed to keep it, although they now have some tighter restrictions placed upon them as far as care and restraints for the animals.

This clause can apply to farms in many ways. If someone's grandfather (or great grandfather or great-great uncle for that matter) homesteaded a plot of land that has a number of buildings and a number of animals — and the family has been actively operating that farm for centuries — a grandfather clause may say that any new laws restricting anything he has on his property don't apply to him. Therefore, if you move into a plot of land next to such an operation, you may be surprised when you can't get a permit to build something that your neighbor has built. He was grandfathered in. You have to abide by the new laws.

Buying the Farm: A Place with Existing Accommodations

You have a lot to consider, question, and weigh when buying a home in the country (many resources can help you with a new home purchase, such as *Home Buying For Dummies,* [Wiley]). Just like buying a house in the city, house shopping in the country can be fun and also frustrating. You may find the perfect plot of land, but the house leaves something to be desired. Does it have the proper number of bedrooms and bathrooms? Do you like the kitchen? Does it have a fireplace? (Hey, you're going to live there, so it should be something you like!) Even if the existing layout isn't exactly perfect, does it have potential for some add-ons or renovations without a huge expenditure of money and time?

So maybe there isn't a fireplace or a big bright kitchen, but the house should come with some basics in proper working order. In an older farmhouse, you may even run up against parts of the structure not being up to code. I'm not talking grandfather clauses here — some things have to be safe, regardless of how they built 'em in the old days. Here's some of what the inspection should include:

- ✔ **The roof and foundation:** Foundations can get cracks threatening the stability of the structure. Roofs don't last forever, and especially in an older house, they may need replacing.

- ✔ **Plumbing:** What's the plumbing like? A hundred years old and in serious need of repair? Do you want to tackle that? Or maybe the owner updated it recently — is she a licensed plumber?

- ✔ **Septic systems:** If the house is on a septic system, does the owner have maintenance records? A septic system gone awry can be a nightmare.

- ✔ **Electricity and wiring:** Older houses may come with older, outdated, not-up-to-code wiring. Bad wiring can lead to fires. Not good.

- ✔ **Heating/ventilation/air conditioning:** How old is the furnace or air conditioner? How about the water heater?

- ✔ **Propane tank:** If your main source of heat is via propane (which is stored in the tank in the backyard), make sure the tank isn't compromised and will serve you for many years to come. Leaky tanks mean lots of wasted propane and larger than normal fuel bills.

- ✔ **The well:** Have an expert test the water for contaminants and check the condition of the well cover and pump.

✔ **Lead:** Older farmhouses may have been painted with lead paint or had lead pipes making up the plumbing system. Lead poisoning is especially dangerous for children; it can cause serious health problems, including lower intelligence.

✔ **Asbestos:** Asbestos used to be used in insulation, floor coverings, and some household appliances until the 1970s, when scientists discovered its harmful side effects. It can break down into microscopic particles that can then be inhaled, causing all sorts of nasty damage to your lungs, including lung diseases and cancer.

You have a lot of questions to ask the current owner when you're considering buying his or her home. But beware — people aren't always as honest as they should be, and they aren't always forthcoming with information you *don't* ask about, either. Hire a home inspector to make sure your prospective dream home isn't a prospective nightmare. Have him or her check the outbuildings as well. Follow the inspector around and ask questions. When finished, the inspector typically points out areas of concern and provides you with a detailed written report.

Starting from Scratch: Planning Your Own Structures

Buying an existing farmhouse or ranch is a lot different from establishing a brand new homestead. Buying something already established means you can move in tomorrow. Building new means you have to navigate a lot of hurdles, and then you have to consider the time involved in building. After looking at the land, you need to consider *where* on that land you're going to live if a house isn't already on it. In this section, I cover some issues involved in starting from scratch.

Don't overlook manufactured homes. Maybe your living requirements are such that a manufactured home perfectly fits the bill. It'll go up much more quickly than a traditional new home, though you may be limited in options and layout.

If moving to the country is your way of going green, remember that new builds are the best time to go energy efficient without much additional cost. Strategically placed fans and vents, windows and sunrooms, good plumbing and insulation, and even landscaping can make your home more environmentally friendly — and cut down on energy costs down the road. (For some ideas, visit www.eere.energy.gov/consumer or check out *Green Building & Remodeling For Dummies* or *Solar Power Your Home For Dummies* [Wiley].)

Dealing with permits and zoning limitations

You can usually build very small outbuildings at will (but check with your local officials to get the particulars). However, really big involved structures require a building permit before you can even dig the first shovelful of dirt.

Each county in each state has its own zoning laws, and you have to check with your city or county before you begin. You'll undoubtedly be restricted on the size and height of your house and barn and also on how many buildings of a certain size you're allowed to erect. For example, if all the neighbors have a ranch-style house, you may not be able to build a two- or three-story house. Or some communities have specific ideas about the outside look of the house, so you may be restricted to building a house with a stucco exterior.

You may have to get a subdivision permit or lot permit. The general contractor typically obtains permits for you, but if you're your own builder, dealing with the hassle is up to you.

If you choose to draw up your own plans, work with an architect to come up with a layout that you like and that makes use of the lay of the land. The plans have to be signed off by both an architect and an engineer, who ensures the structures will stand up to things such as wind strength and snow load.

After the house plans are complete, several tests need to be done on the land, and several things need to be okayed before the almighty permit is granted:

- ✔ If you won't be on city water, you need to build your well before anything combustible can be erected (the well serves as the fire-suppression water source). The water in the well also has to be tested to ensure that it's drinkable.

- ✔ If you'll be putting in a septic system, you have to get a percolation test to see whether the drainage of the soil can support the septic system.

- ✔ You need to hire a surveyor to map out the official property lines. The officials have to okay where on the land the building will be.

This is by no means an exhaustive list. Your county has its own rules, and you may have to make adjustments that people in other parts of the country do not, such as specific precautions for houses in a flood plain or in Tornado Alley.

After the county approves your plans, you get to wait for the wheels of government to turn. For my house, it took 13 months (including all the testing). A friend had to wait 17 months. It's a frustrating time, my friend.

County, may I? Experiencing the permit process

Getting the building permit for our house took us 13 months. We found that, as an owner/general contractor, getting sign-offs was a lot harder than it would've been for a professional general contractor. But we saved tens of thousand of dollars in doing it ourselves. We visited the county office more times than I can count before we finally walked away with the official go-ahead.

As for the barn, we first petitioned for a permit while we weren't in the boundaries of any city — we were officially an unincorporated county area. Before we nailed down a builder (building the house was an experience that, although ultimately a good one, was not one

Hubby ever wanted to do again), the area was annexed into the nearest city. Guess what? We had to go through the permit process again! And pay the fees again. Aaaarrrgh! Fortunately, the city we're now a part of was pretty good to work with, and getting the barn permit didn't cost too much in money or headaches.

As for small buildings, we've built several for animal shelters and various storage units. We didn't have to even apply for a permit if the building was less than a certain square footage (this number varies widely — check with your county officials). We even erected an 18'-x-12' Tuff Shed and didn't need a permit because it was under the permit-required limit.

Permits are a hassle, but you have to live with them. In the long run, they're in place for the betterment of the community — after all, you don't want somebody to move in next to you and build a factory that'll bring a lot of smog and traffic that you didn't anticipate in your wildest nightmares!

Sometimes you can find ways to get around the rules without hurting anybody and at the same time making your new place your dream place. For instance, some friends bought a subdivided portion of land from a homesteader who needed some money. At the time they bought the property, they had no idea they wouldn't be able to build a house on it. They assumed they *would* be able to! But when they tried to get a permit, they were told they could only build a barn (not a permanent human residence). They got around the law by building a barn with the minimum requirements of living — there's no phone or indoor plumbing!

Getting the almighty Certificate of Occupancy (C.O.)

A *Certificate of Occupancy* means the officials have come in and inspected the building and determined that it's up to code and safe for occupancy.

Before your house is completely built and an official inspector has given you the stamp of approval, you can't officially live in your newly built home. Doing so, if you're caught, results in fines and worse — you could be forced to move out until you do get the stamp of approval.

This can be really rough, especially when you've sold your other home and are living in temporary quarters, and your new place is almost done. You really just want to move in. The plumbing is done and the roof is intact, but maybe not all the drywall is up or all the trim is finished. Oh, how annoying that can be! It's so close! To your sensibilities, there's no reason you can't move in and stop paying the temporary rent at the motel. You could be saving so much money if you could just move in!

Yes, we were tempted. We sold two houses (the marriage came with two households, and after consideration, we decided neither was suitable for what we wanted to do now that we had more options). Fortunately for us, we were able to start the building process before everything else was sold, so we had some leeway. By the time the second house sold, the new house was *almost* ready to move into. We were fortunate to have a friend who let us use a trailer for the two months before the new house was finished and the C.O. was granted and we officially moved in.

Chapter 4

Getting Along with Your Neighbors

● ●

In This Chapter

▶ Meeting and greeting

▶ Considering what the neighbors think

▶ Obeying local ordinances

▶ Not messing with the wildlife

● ●

*I*n the country, the distance between houses doesn't always equate to distance in relationships. Rural residents often share ties to the land, a love for a slower-paced life, and perhaps generations of common history. They help each other out, and they want to make sure that their community remains a pleasant place to live.

In this chapter, you find out how to be a good neighbor. I bring up some points you may not have considered, such as respecting what types of operations were there before you and dealing with land issues. I also explain how you can relate to your furry and feathered neighbors, the wildlife that may give you a visit.

Opening the Lines of Communication

Getting to know your neighbors on a first-name basis and becoming friendly with them is a good way to stave off tense future confrontations. If you know what makes them tick (and they know you as well), it's easier *not* to do those things that offend people whom you had no idea those actions would offend.

In some cases, you can discuss big upcoming projects with your nearby neighbors and ask for their suggestions. If they've already been through a similar project, they may have ideas on how to get it done with the least amount of inconvenience to everyone around. You don't have to ask permission, but if you know your neighbors and some of the things that annoy them, you can take small steps to be considerate of their situations.

If something potentially volatile does come up, casually meet face-to-face and discuss your concerns instead of keeping everything in until somebody blows up. If you're on a friendly basis, confrontations can be a little easier. Here are some tips:

- ✔ Don't place blame, and give people the benefit of the doubt. Instead of saying "Quit driving so fast!" you can say, "I don't know if you realize it, but when you drive fast on this road, it really kicks up a lot of dust."

- ✔ Pick the right time to approach the person, when your emotions won't get the best of you. Don't wait until you're furious. Be nice, and try to find a solution together.

- ✔ If you can't come up with a solution on your own, suggest looking for outside help. Perhaps you can say, "Maybe we can ask the county to spray the roads more often." Or suggest calling the local extension service for suggestions.

Beyond just your immediate neighbors, be friendly with the people you pass on the street. Volunteer for community events and just get involved. Not only will life be more fun, but you'll also be better able to work things out with other people in the community in a friendly, cooperative manner.

Appreciating the Community as It Is

More likely than not, the community you decide to move into will be running smoothly with all sorts of personalities and operations working together. There'll undoubtedly be conflicts at times, but on the whole, the community will be doing well, and it's in your best interest to go with the flow of the current lifestyle, not only so you fit in but also so you thrive.

Do your best to embrace the community and try to fit in. It'll make your life — as well as the lives of everyone else in the area — that much better. Here are some things to consider:

- ✔ **Current operations:** When you're deciding where to live, consider what types of farms are already in the area or what kinds of other operations are around. Try to find an area where the types of things already going on are the types of things you can live with — and that will be compatible with what you intend to do with your own property.

- ✔ **Legal restrictions:** Be sure the local laws permit what you dream about for your life on the farm. You may run up against zoning laws, noise restrictions, and wildlife laws. Before you decide to embark on an endeavor, check with your city, county, and state (a good place to start is on your state's Web site, typically www.<your state>.gov) and see what's allowed. You can also call the various offices for questions you can't find answers to online.

Going with the flow after an unsettling development

Everyone is connected, so ethically, owning land doesn't give you permission to do anything you want with it. Really think about how what you're planning to put on your property will affect your neighbors.

A lot of development is going on around the country. Babies keep being born, growing up, and then needing places to live and raise their own babies. I've been personally affected by major development coming into an area that, when I bought the land, was designated as *greenspace* — a section of land intended to remain "green" with grass or brown with sand or whatever color nature gave it. All my neighbors lived on at least 5 acres (most actually had two lots, or 10 acres). But somebody with enough money persuaded the city council to change the zoning, and now there are several hundred new homes, all on 1/4-acre lots plopped down in the midst of the rural area that I and my neighbors moved to with the hopes of retiring in the country.

This new neighbor (the developer) didn't care about the area or how the changes would affect the existing residents. Even though we fought for more than a year to stop the development, big money won out. A better behaved "neighbor" would've listened to the concerns of the residents and altered the plans. That didn't happen.

Somewhere along the way, we all realized that the development was going in regardless of our efforts. We asked whether he'd consider making the lots at least 2½ acres. We got laughed at. We then wanted at least 1 acre and got another resounding *no*. The developer's refusal to be flexible and the city council's lack of vision were frustrating, and now this once-tranquil place is no longer recognizable.

This sort of situation isn't always foreseeable. The best thing to do is to be able to take things as they come. My neighbors mostly moved away, but I opted to stay because I still like the location despite being a little crowded in. I still have land and animals in a location that works for me. What I have lost is the open air feeling, with all those houses so close, but I'm staying put.

- ✔ **The way things work:** Moving in and trying to change everything — whether you're introducing something new or trying to get the existing operations to stop doing something — won't make you very popular (unless, of course, those changes benefit everybody and everybody agrees with them). For instance, storming the city council meetings and demanding that Farmer Joe get rid of his cows because you don't like the smell will get you nowhere fast, but storming that same meeting and demanding a traffic light at a certain dangerous intersection will likely be met with positive results.

- ✔ **The atmosphere of the place:** Remember your reasons for moving out into the country. Chances are one of those reasons has something to do with a quieter, slower, simpler lifestyle. Making a big stink (maybe literally) is not how to win friends and influence people.

In this section, I discuss how you can work with current operations and laws.

First arrivals: Accepting what's already there

Be considerate of the surrounding homes when you move in and want to start up an operation. Keep in mind that when you move in, whatever is already there was, indeed, there first, and you don't have a right to expect an existing operation to stop doing business just because you don't like it.

I know of some city slickers who moved to the country to live the simpler life. They found some great, inexpensive property that was far enough out there to be "country" but still relatively close to civilization. One minor problem was the *rendering plant* (a meat and byproduct processing plant where operators try to use every part of the animal and turn it into something useful, not necessarily food). They built a house and got a few animals. After a few months on the property, they decided the incessant smell coming from the plant was unbearable and tried to get the plant to stop. Of course, they lost the battle and eventually moved.

If you move next to a large pig or dairy farm and later decide you can't take it, that's your problem, not the problem of those who came before you, so choose your location carefully. (For tips on choosing your property, see Chapter 3.)

Everybody else is doing it: Acknowledging grandfathered-in implications

Operations that are protected under a *grandfather clause* can keep doing business as usual when everyone around them can't legally do the same thing (see Chapter 3 for details). Older established farms that have been doing certain things for years may be *grandfathered in,* or allowed to do something that current or recent city councils have outlawed for future residents.

For example, look at zoning laws. Say Farmer Joe has 1 acre, and his land contains a large barn and dozens of animals. You buy an acre right next to Farmer Joe, but you're not allowed to build a big barn, and you're restricted to only two animals. "Not fair!" you say. Sorry, but that's what the grandfather clause does. And even if Farmer Joe decides someday to sell his land, the rights that he had to do business as usual may or may not transfer to the new owner — that all depends on your local laws.

One other implication to this clause is a sort of use-it-or-lose-it issue. If Farmer Joe is allowed to have ten llamas (and his neighbors are allowed only two) and his herd eventually dwindles down to only a few animals, he better increase his herd within a certain time period, or he'll lose his ten-llama limit.

Just because your neighbor is doing something, that doesn't mean you'll be allowed to do it, too. When you buy your land, make sure you have an idea of what you want to do with it so you can be sure the current zoning laws will allow it.

Objections to Certain Farms . . . or What's on Them

As a hobby farmer, you probably aren't interested in overdeveloping your land (trying to change the face of the landscape by putting high-density housing right in the middle of your rural neighborhood). That's just not what you moved out in the country to do. But you can engage in dozens of other bad behaviors that are just as annoying to the existing residents.

If you have gobs of acreage, these issues probably won't come into play. You'll live far enough from your nearest neighbor (and will have a sufficient buffer zone) that nothing you do will reach anyone else. But if your plot is only an acre or two and your neighbors are close by, please take their feelings into consideration before you embark on a venture that could make them feel not-so-neighborly. This section covers some of the major areas of contention.

No matter how close or far away you are, keep your operations clean and in good working order. A sanitary farm minimizes odors and is less likely to spread disease, unleash a plague of rodents (such as rats or wood mice), or contaminate the groundwater.

Turning up the volume

Remember how living in the city sometimes meant having to listen to your wild neighbors' parties going on at all hours of the night? Well, some critters and machinery on your farm can also cause very unwelcome noise pollution.

Animals that raise a ruckus

Filling your pastures with animals that keep the neighbors up at night or cause continuous raucous rackets during the day is not very nice. All the animals in this list are common farm animals, but each can be very noisy. Depending on the lay of the land, your neighbors may be able to hear your animals up to a mile away. Here are some of the biggest culprits:

✔ **Roosters:** I think no other sound says *farm* more than a rooster crowing at dawn. Some people (me, me) love to hear sounds like that. But not everybody does — especially your neighbor who works till 2 a.m., comes home, and just barely gets to sleep before your rooster starts letting him know that the sun is on the rise.

You don't have to have a rooster to get eggs from your chickens. Chickens lay eggs regardless of the proximity of a rooster. Having a rooster around only means those eggs can get fertilized and possibly turn into baby chickens. Unfertilized eggs are going to happen without him.

✔ **Peacocks, turkeys, geese, and guinea fowl:** Most birds can get noisy. This makes them great alarm animals (see Chapter 9 for more on alarm animals), but it can make them very bad neighbors. If your farm is close to other houses, starting up your own aviary isn't a great idea.

✔ **Camelids:** Male llamas and alpacas fight if their baby-making equipment remains in working condition, regardless of whether females are around. It's part of establishing who's in charge. The screeches they make probably aren't like anything you've ever heard. The neighbors, if they haven't heard it before, may think something's terribly wrong over on your property — and if they have heard it before, they may get to thinking that you're a terrible, neglectful owner.

Llamas and alpacas also make unusual noises when something new comes into their world. When I introduced goats to our flock, the llamas made this weird screech-hum for a few days until they got used to the idea of those funny-looking, very small new "llamas." (See Chapter 9 for more information on camelids and why they make noises.)

✔ **Dogs:** Some dogs just like to bark. Some bark because of separation anxiety, and others bark only when dogs, deer, or other foreign creatures come into their territory. But keeping a dog who has a barking problem can cause rifts between neighbors.

Some dogs can be trained not to bark, just like they can be trained to do anything else. It's more of a personality question than a breed issue; some dogs whose breedmates can be beautifully trained never really learn what *no* means.

✔ **Goats and sheep:** Baa, baa, bleat, bleat. These guys can be noisy, especially when they see humans coming toward them, indicating that food is forthcoming. These animals are the least noisy of the ones I've encountered, but you never know what's going to set off a cantankerous neighbor.

Stopping the noise isn't always easy (or possible), and chances are pretty good that your neighbors will have animals, too, and will be more understanding than your former city neighbors would be. But just be considerate of the noise level and the proximity to your neighbors' existing houses.

Of course, not all animals are noisy. Rabbits, worms (silk- or earth-), and fish, for instance, aren't likely to cause your neighbor to pound on your door in the middle of the night asking you to keep them quiet!

Noisy machinery

Animals aren't the only ones who can cause annoying noises. When most farm machinery is in operation, there's usually an accompanying racket. Tractors, lawnmowers, snowblowers, and the like all produce high decibels. Most of this can't be helped — you have to do your job (as do the neighbors), and these machines are what you need to use to get that job done. But be considerate and limit their use to daylight hours if possible.

The exception to this rule occurs during harvest season. Farmers are going to begin at the crack of dawn and extend as late into the evening as possible, noise or no noise. Their timeframe to get in the crop or lose out on a whole year's income is short. If you're doing the harvesting, or if you're enduring the noise from your neighbors, the noise is something that you — or they — just have to deal with.

With the exception of harvest season, the crack of dawn is not an acceptable daylight hour!

Ditto on the be-considerate advice with running your generator. Yes, you may sometimes have to run the thing at night (power outages during a blizzard or subzero temperatures), but for everyday activities, limit the noise to daylight hours.

Raising a stink

What's that smell? Farms inevitably produce some odors that aren't particularly, well, fragrant. Because a hobby farm is typically operating on a relatively small scale, those aromas may not get to a gut-wrenching level. But odors your farm produces still may be something to think about restricting:

- ✔ **Silage, haylage, and baylage:** These mixtures are made from different recipes, but all are fermented and later used as feed. *Silage* is a mixture of crops (such as corn, grass, or oats) that's cut up into small pieces and put into a silo or pit and allowed to ferment. *Haylage* and *baylage* are made of hay and grass. (The process for making baylage is simpler and is more typical on smaller farms.) This type of feed is high in the nutrients that promote milk production and is thus the feed of choice for dairy farms, but the stuff really stinks — it's fermented, after all. As in most other stinky situations, the small farm probably won't produce enough odor to make the neighbors raise a stink, but you may have to put up with a large-farm neighbor who is.

- ✔ **Large animals:** Large animals produce large amounts of stinky stuff. This is just a fact of life. If your land is very close to nonfarm land, the neighbors may not appreciate your animals the way you do.

✔ **Pigs:** I don't mean to pick on pigs. If I were able, I'd have a pig or two on my farm. (Hubby says a resounding *no* because he worked with the critters in vet school and won't do it again.) But they can really be stinky — so stinky, in fact, that I think they may be a great deterrent for those new housing developments people are always trying to build in the country!

Seriously, a pig or two probably won't bother anybody, but if you have a large herd in an area that doesn't support it (for instance, your neighbors are very close by), you may have the local law enforcement knocking on your door at some point.

Besides the pig's back-end products and the food (you can feed stinky garbage to the undiscriminating eaters), pigs attract flies, which can be very bothersome.

✔ **Birds:** Chicken farms can rival pig farms for the winner of "most malodorous." But if the farms are well maintained (you pay close attention to cleaning), the odor can be acceptable. Small farms usually don't produce enough of that odor to become a nuisance.

Some people swear that duck pens are less stinky than chicken coops and instead keep the ducks for eggs.

You're not completely off the hook if you choose not to keep animals, either. Using manures to enhance the nutrients given to your crops is good for the crops, but it can really make the air stinky. (See Chapter 18 for more information on using manures in planting.)

Ratcheting up the danger

Threat of danger to you, your animals, or your guests (invited or not) comes along with your wide open property and the animals, tools, machinery and even structures you acquire. It's a sad state of this day and age that Americans are so litigious. Sometimes it's deserved (for instance, you deliberately neglected to fix your fence and your bull got out and destroyed your neighbors' garden and killed their dog), but many times, accidents happen despite your careful attention to safety. Although this shouldn't scare you off from having a farm, it should make you aware that injuries can occur. Be careful, and consider beefing up your insurance.

Potentially dangerous animals

Any animal is potentially dangerous. You can find reports of people being attacked by sheep, of all animals! But a few animals have gained a reputation for being especially risky. Keep your neighbors in mind when you think about the following critters:

✔ **Bees:** Bees don't stay put in the nice hives you made for them. People can stumble onto a hive in your field (no, they shouldn't be trespassing, anyway!) and stir up the hive and thus get multiple stings. Or a person may just unwarily anger some of your foraging bees. As with any potentially dangerous situation, if a person is injured (specifically, if an allergic person is stung by your bees and ends up in the hospital), you can end up with a lawsuit on your hands.

Extra precautions you can take include keeping your hives as far away from occupied buildings and property lines as you can (put them in the middle of your fields) and limiting the number of hives you have.

✔ **Exotic species:** Some animals, especially the more exotic ones such as llamas or buffalo, may be more apt to attract kids to come over to the fence to see just what that critter is. If you're not able to keep watch 24/7, make sure the fencing is secure (so animals can't get out and curious kids can't get in) to prevent injuries or worse.

✔ **Intact males and new mothers:** When mating time rolls around, the males get aggressive. Also, mothers with newborns are very protective and can attack anyone who gets too close to their young ones. During these times, keep special watch on your animals.

✔ **Any livestock on the roads:** If any type of livestock gets out of the fence and ends up wandering on the road, this is a recipe for disaster! Car versus animal can mean a serious accident, harming valuable animals as well as humans. Keep those fences secure.

Potentially dangerous equipment and substances

Animals aren't the only sources of danger on the farm. Tractors, grain bins, sharp tools, and ponds have the potential to inflict serious injury or even death. Take special care of the following:

✔ **Machinery on the roads:** Driving large machinery down the road is necessary in rural areas, particularly during harvest time, so patience and understanding are important — as is safety when driving around these vehicles. If you're driving your machinery at less than 25 mph, make sure you've affixed a slow-moving vehicle sign (which looks like an orange triangle) to the back.

✔ **Tractors:** These large moving vehicles can run over somebody, or their attachments can hit someone and cause serious cuts, broken bones, or concussions. Stay in control while driving the tractor, and be aware of what's going on around you.

✔ **Grain bins:** These seemingly innocuous storage facilities are responsible for many deaths each year. Someone can fall in and get buried alive, suffocating before you even know there's trouble. Devise a buddy system anytime you need to work with the grain bin.

✔ **Sharp tools or power tools:** Even when these tools are in the right hands, accidents happen, sometimes cutting off fingers or puncturing bodies. Keep your workspace clear, and never use these tools when you're distracted or under the influence of some mood-altering substance, even prescription medications.

✔ **Ponds:** People, especially small children, can fall in and drown in ponds very quickly. Yards with ponds should be fenced in. If you aren't using the pond in the middle of a field to water your animals, you can even put a fence around the pond itself.

✔ **Pesticides, insecticides, and herbicides:** These chemicals are formulated to kill pests, but they're dangerous to others as well. Keep them well labeled and out of reach of curious children or animals.

Lighting up the sky (and your neighbor's house)

Sometimes you can't help shining a light in a direction that may annoy someone — for instance, when you're working into the late hours during harvest season, or if an animal has broken out of the fence during the night. But for those occasions when you can control the light, be courteous in your usage and placement of these potential nuisances.

Not only does excessive light interfere with the view of stars, meteor showers, and other highlights of the night sky, but it may also interrupt your neighbor's sleep. Who wants to have a neighbor's lights shining in their bedroom windows through the night? Position your lights so they don't spotlight someone else's house.

Also keep lights in mind if you have to do some emergency night work, especially if you need to run some activity all night long or if you're having a special occasion that lasts into the night. Just common sense stuff, here.

Handling Land Issues

If you settle on undeveloped land, some clearing may be in order to make it suitable for building. At the least, you'll probably be clearing some brush, but you may also have to get rid of a tree or two or clear a path for a road. In this section, I discuss land-clearing issues and easement.

Keeping the trees

I used to live next to a beautiful wooded area that was a local favorite for hiking. No longer. When a developer moved in and convinced the city council to change the zoning laws, all the trees were felled, and now there are hundreds of houses with bare backyards.

In this day and age of disappearing greenspace, preserving some of the plants that are on your property is a really good idea. Yes, some trees just need to go, but try to plan the landscape around some of the plant life that's native to your area. Native plants are a good choice because the soil and climate are right for them to thrive. A properly planted and established native tree requires very little long-term maintenance. These plants help preserve the particular environment they live in, providing food and shelter for native animals.

The benefits of woods and noncrop vegetation aren't just aesthetic. Trees and shrubs can prevent erosion, filter pesticides and fertilizers from the soil before they enter streams, provide windbreaks to protect crops and livestock (and block odors), provide wood and fodder, increase biodiversity, and more. You can visit www.unl.edu/nac for more info on agroforestry and the relationship between farmland and trees.

Allowing access

You may encounter the issue of allowing access to your property or someone else's (called *easement*). Shared property lines can sometimes mean the owners of the respective properties have very different ideas for the land. For instance, you may decide to sell a portion of your land, but the only way for the new owner to get to it requires a road to be built right through the property you didn't sell.

You can also come up against public-access rights to streams, for instance. I know of a couple who bought some land that had a stream running through it. They had to agree to allow fishermen to access the stream, which meant they had to agree to allow anybody to walk through their property at any time. Conditions of the sale make it impossible for them to erect gates (maybe to keep animals in) that would prevent that access. This is just another thing to look out for when you're evaluating property to acquire.

Dealing with the Local Wildlife

Sitting on your back porch and watching deer in the fields is a sight unseen in the city, and it can give you a sense of really being in the country. You may find, though, that some wild animals aren't welcome on your property, and getting rid of them may not be as easy as you thought.

The first time a llama was born on my farm, I was just so excited to see a new life. A few weeks after the birth, I noticed a coyote perusing the grounds. (I just love to hear them call to each other in the night!) I excitedly told my hubby we had this canine visitor. He brought me back to my senses by telling me that the visitor was visiting because we had fresh meat to feast on. I was horrified, both at my naiveté and at the fact that my new baby was at risk.

This section explains a bit about taking care of your wildlife problems responsibly.

ID-ing the usual suspects

The list of wildlife pests runs the gamut from rodents to raccoons, from coyotes to herons.

Like it or not, deer love to come onto people's property and munch on anything available. In the wintertime, deer are going to try to get at your hay pile unless it's protected. (At least cover it with a tarp; at best, store it inside a barn.) And during the growing season, deer often wander into gardens and sample the fresh salad. (You may not even get to benefit from the fruits of your labor.) High fences and other deterrents are about the only thing you can legally do to control them.

Coyotes — creatures with big teeth who love to snack on poultry and baby animals — are prevalent in many areas across the U.S. This is one reason you may need a guard animal — one big enough and territorial enough to fend off these critters and discourage them from returning.

Adhering to wildlife-protection laws

Often, people are tempted to immediately go after offending wildlife with poison, shotguns, or traps, but this isn't advisable. Small rodents that are digging holes in your yard or even woodpeckers that are putting holes in your wooden structures may have laws protecting them, and you may or may not be able to remove those animals yourself. I've had problems with flickers' (a form of woodpecker) putting holes in my house. The only legal thing I could do was catch them in a live trap and take them to the state officials for them to release them elsewhere. Pretty annoying!

A fine example of a beautiful beast that turned out to be a whopping pest is a case of the great blue heron versus an Indiana fish farm. Herons are protected birds in Indiana. But Farmer Bob had a big pond where he raised fish that he sold to the local market. The herons liked fish and discovered that Farmer Bob's pond was an easy source of food. The state laws said he couldn't shoot, poison, or otherwise do anything to harm the birds. What a dilemma! So Farmer Bob erected a "ceiling" of netting to go over the pond so the birds could no longer get in. It was an effective solution, and no herons were hurt in the process.

If a pest appears and starts wreaking havoc, check with your state before taking matters into your own hands and risking heavy fines or even jail time. Contact your state's department of wildlife resources or maybe even your local extension service (www.csrees.usda.gov/Extension) to see what you can and can't do.

Looking for creative (and harmless) pest control

If the local wildlife is wreaking havoc on your farm, look for a creative ways to thwart the critters without hurting them. Here are some of your best bets:

- **Physical barriers:** These barriers may include trees, rocks, and ravines. See Chapter 8 for information on physical barriers.

- **Spikes on eaves:** To keep birds such as woodpeckers from roosting on (and pooping on) the eaves and nooks and crannies on your roof, you can install metal spikes to deny them a comfortable place to land.

- **Electric fences:** These fences can not only keep your animals inside but also deter some predators from coming in.

- **Guard animals:** See Chapter 9 for information on guard animals and some ideas of which animals fit the bill.

- **Humane traps:** Capture the pest in a live trap and either release the animal yourself in a recommended area (ask your state's department of wildlife resources or your local extension service for suggestions) or take the animal to those agencies and let them deal with it. We use live traps for raccoons and skunks.

- **Nontoxic products with an unpleasant taste:** Hot sauce, chicken eggs, or commercial substances such as Ropel may discourage pesky visitors without actually harming them.

- **Commercial products made from coyote urine:** These substances keep other, territorial coyotes away. Mark your own territory in the same way territories are established in nature.

- **Ultrasonic devices:** These devices make a high-frequency sound that humans can't hear but that are very annoying to animals.

- **Scarecrows and fake owls:** These plastic or plaster critters can fool some smaller animals (especially rodents or small birds) into thinking the area is being patrolled and will thus stay away.

You can also ask around for other ideas. What do your neighbors do? What does the department of wildlife resources or the extension service suggest?

Keeping local wildlife and exotics on your farm

Some animals that would otherwise be just fine in the wild somewhere may thrive on your farm. However, wildlife agencies strongly discourage keeping wild animals, so most wildlife come with laws restricting your handling of these animals — let alone your keeping them.

I've heard of some people trying to keep cougars (mountain lions) as pets, without taking into consideration local laws against keeping such animals or neighbors who may fear for their children or livestock. There have also been cases where farmers deliberately overfed wild ducks (the ducks became too heavy to fly away) in order to keep them on their properties for eggs or for "controlled hunting" purposes — also against the law in many states.

But there are exceptions about which animals you can raise. I know farms that keep buffalo, antelope, musk oxen, and reindeer. (And I've always thought it'd be cool to get a couple of kangaroos, but I haven't investigated whether it's possible). Not only are these animals very cool, but they can also produce some wonderful products, such as the very fine fiber of the musk ox or the regularly shed antlers from the reindeer.

Although having a lot of land way out in the country may seem like a perfect opportunity to get an exotic, formerly-wild critter, you really need to contact your state's wildlife department to see which animals you *can* have and what you can do with them. And certainly don't simply go into the woods and try to catch an animal to take home. Ignoring these laws can get you in trouble with both your neighbors and with the law (large fines and even jail time).

Part II
Down on the Farm: Getting Your Property in Order

The 5th Wave   By Rich Tennant

The pH level in the soil seems a bit high.

In this part . . .

Maintaining and living in a country farmhouse isn't like living in a city apartment or even a house in the suburbs. You need some outbuildings and special equipment to keep things moving smoothly, and you need shelters for animals and storage for small tools and large equipment. You'll probably have to build some of your own roads, and you're likely to need some alternative power sources. This part helps you make sure your operation has what it needs to prosper.

Chapter 5

Getting Out of the House: Outbuildings and Enclosures

- -

In This Chapter

▶ Places to store things

▶ Places to house things

▶ Places to grow things

- -

Your property needs many structures besides a house in order to make life on the farm successful. You need shelters for your animals; a place to store food, farm equipment, and supplies; and depending on what you intend to do on your farm, a variety of other enclosures or special areas. In this chapter, I describe some of the buildings and enclosures that tend to crop up on hobby farms.

Raising Barns

What's a farm without a barn? Barns are multipurpose structures that can serve as shelter for your prize racehorses, a storage facility for hay or farm equipment, or a place to hold an old-fashioned barn dance. You can use them as workshops or garages or even convert them to living quarters for your favorite mother-in-law. They come in many designs, and each has its benefits and drawbacks.

What you plan to do with your barn helps determine how big it needs to be, the layout, the building site, how many windows you want for ventilation, and other considerations. Aesthetics also plays a part in what you build because the barn contributes to the overall look of your property. Your budget, unfortunately, also has a big say as to the final barn building.

Barns can be temporary or permanent. They can be made out of steel, wood, aluminum, canvas, brick — pretty much any building material. They can be any size or shape. But in the long run, a barn is really a necessity, regardless of the particulars. In this section, I discuss some common barn styles.

Be sure to build your barn to code, especially if you'll be including electricity, plumbing, and an emergency phone. And don't forget maintenance — you'll probably want to paint or stain the barn every three to five years, keep the doors and gates in working order, and repair any rodent chewing.

Traditional structures

The traditional barn style, probably the image you get when you think of a barn, has a *gambrel* roof (it looks rounded, but it actually has four slopes). The good thing about these traditional structures is that they can provide good shelter from the elements because of their closed environment — four walls and a solid roof keep the weather out, and whatever's inside stays warm and dry. Besides being aesthetically pleasing with a nostalgic look, traditional barns are functional. You can keep animals in the lower levels and hay in the upper levels — simply rig up a system of pulleys and clamps to easily haul the hay up there.

Because they're so solidly built, you need to put a little more time and money into erecting these structures, but you can find so many plans and ideas on how to organize the space. Check out www.barnplans.com for some ideas on what's out there.

Post-frame construction: Pole barns

Pole barns use a building technique called *post-frame construction;* its key feature is that the main supports for the structure are widely spaced beams or poles, usually wood that's pressure treated to reduce any rot (see Figure 5-1). These buildings are simple structures and can be wall-less (just poles supporting a roof), or you can add any sort of materials for walls, even aluminum. You can have pole barns built at half the cost and half the time of the traditional structures.

Check out some of the books on the market (such as *Practical Pole Building Construction: With Plans for Barns, Cabins, & Outbuildings,* by Leigh Seddon) or Web sites such as www.easybuildings.com to see how it's done and what to buy to do the building.

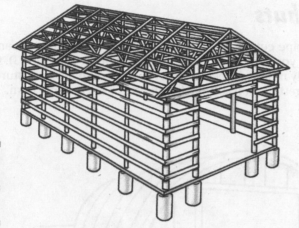

Figure 5-1:
The skeletal
structure of
a pole barn.

Lean-tos

A *lean-to* (also called a *run-in shelter*) is typically a three-sided structure with a very slanted roof (see Figure 5-2). The lean-to can be standalone, or it can lean up against another building that functions as one of the structure's three sides. You can use lean-tos to store hay, equipment, and so on, or it can function as an animal shelter.

These structures are typically smaller than a traditional barn, but they can be much less expensive to erect. They usually don't have any sort of wall or door on the front side, so they're not the best for complete protection from the elements; however, they may be suitable shelters for llamas and other hardy pasture animals.

Figure 5-2:
A lean-to
provides a
roof and
shelter on
three sides.

Quonset huts

Picture a giant pipe cut in half, leaving a shape with a flat bottom and a domed top, and you have the age-old Quonset hut (see Figure 5-3). Quonset huts are typically made of steel, are easy to erect, and are very sturdy, but they're not as pretty as a traditional structure or even a pole barn. They're more utilitarian.

Figure 5-3:
A Quonset
hut.

Choosing Animal Enclosures

Besides barns, which can serve as enclosures for large or small animals, you may be interested in some smaller, less-involved enclosures. This section runs through some of the ways you may want to give your animals a home.

Going predator-proof: Small animal enclosures

You can successfully house some smaller animals (chickens, rabbits, and so on) in smaller shelters — no need for something as big and involved as a barn for them.

Mowing with the chicken tractor

You can use chickens as land-clearing devices! My farm devised an enclosure called a *chicken tractor* (at the suggestion of a friend — I can't claim the patent). This mobile pen is simple to make. It has four sides and a roof, but the bottom is open. Use chicken wire on all sides over some sort of frame. The frame should be of a material such as 2" x 2" pieces of wood — something that's stabilized for UV light (we first tried using PVC pipe, and after the first season, the pipe became brittle and fell apart). It should be high enough and wide and long enough to comfortably allow several chickens to stand underneath it.

After you make this device, place it in an area you want cleared. Gather a few chickens, put them under the tractor, and then let them have at the grass and weeds underneath them. After a few hours, move the structure (and chickens) to another area. Before you know it, your lawn has been mowed!

We call this device a *tractor* rather than a *mower* because the chickens not only chew on the grass but also scratch up the soil, dig up bare weed seeds, and turn in those weeds and the inevitable chicken poo. If your soil isn't rock hard, all that scratching and pecking can turn up a couple of inches worth of weed seeds and bugs and other things.

Chicken coops

Not even free-range chickens want to spend all their time roaming the countryside (well, your yard). Chickens need a place to go at night, more to protect them from predators like coyotes, foxes, skunks, or mountain lions than anything else. These structures range from simple to very deluxe, but the basic components are four walls, a roof, and either a floor or a means of having the walls close enough to the ground that vermin can't crawl under and get inside (though a floor is really ideal).

Coops come in many styles. A friend of mine even fashioned a chicken coop out of an old camper; he cut a few openings and attached doors on hinges with locking latches so he could reach in and gather eggs or deposit food and water. The coop that sits on my property is a more traditional structure, made of wood with four sides, a floor, a roof with shingles, and small doors for egg collecting and food delivery, with a big chicken wire-panel on one wall for ventilation. Your imagination is your limit on design.

Rabbit hutches

Rabbits, like other animals, need protection from predators, and the very dogs that you use to guard other animals may be the same critters that threaten the rabbits. Outdoor rabbit houses, called *hutches,* are typically aboveground, secured shelters, so dogs (and skunks, raccoons, and coyotes) can't even get close to them.

The bottoms of the cages are typically a wire mesh-type material so the frequently produced poop can flow freely out of the cage. The hutches my dad used had chicken wire on three sides and on the bottom, and the other side and the roof were wood. Such wire-and-wood hutches can adequately protect the rabbits from the weather (rain, snow, and so on); for warmth, you can put some straw or other bedding materials inside.

Keep the hutches inside fences where predators can't get close. When rabbits get nervous and freak out, they kick. The kicking feet can go down through the wire mesh, damaging the rabbits' feet and legs. They can actually wear their hocks down to nothing (a *hock* is like an elbow on the back of the leg). This can lead to bone infections that are nasty to deal with and are pretty much fatal.

Braving the elements: Medium-to large-animal considerations

Some medium to large animals don't necessarily need to be in a barn, but they still need or appreciate some sort of shelter, even if it's very rudimentary. You may also want special enclosures for setting up couples during mating season — and facilities for handling the subsequent births. And of course, you need somewhere to quarantine animals who end up on sick bay. Read about some of these less-intensive structures here.

Run-in shelters

Goats, sheep, cattle, and llamas are generally okay in the elements. In fact, for the first few years on our farm — even though I felt bad about it — we didn't yet have any protection for the animals, other than the trees and shrubs. I was astounded at how the llamas, which are native to the Peruvian Andes and are very rugged individuals, didn't seem even to notice that bad weather was happening; they just sat there, seemingly unaffected by the sideways, howling wind and unending snow. (Actually, some untrained llamas are leery of going into any sort of shelter! You need to teach them that it's okay to come in from the storm.)

Although providing shelter isn't a life-or-death situation in many cases, it's a nice thing to do, and it'll most likely extend the life of your animals. I suggest having at least something to block the wind for all animals.

Besides the barn, we've erected a few other less-involved shelters, including the following. Here are some ideas you can use:

- ✔ An otherwise unusable shipping crate, put into the goat pen, has been adopted wholeheartedly during inclement weather — you won't believe how many animals can fit into such a small space!

- ✔ A plastic dog kennel makes for a great shelter for a newly weaned or yearling animal (such as a sheep or goat). Line it with some raw fleece, using the sheared stuff that's too matted or otherwise not able to be processed into spinnable wool.

- ✔ You can always make a lean-to out of plywood — a temporary shelter with three sides and a slanting roof. We made such a structure, and even the timid llamas didn't mind going into it when a nasty wind, rain, or snowstorm visits the farm.

Birthing facilities and nurseries

When a female is about to give birth, you want to take her into an area where she isn't stressed by fellow animals vying for food or attention — or by predators. This may be just a room in the barn, or it can be a whole separate structure, but in any case, it has to be an area that separates the new-to-be mother from general activity. (For more on animal births, see Chapter 12.)

After the birth, some newborn animals need to stay inside (with their mothers, of course) until they're tough enough to deal with the elements. Kittens, puppies, and so on need special care until they're developed enough to manage on their own. Pigs can't regulate their body temperatures in those first few days, so they need to be kept indoors and warm; a birthing house for pigs, called a *farrowing house,* needs to be temperature controlled.

Besides puppies, kittens, and pigs, horses should also be housed. Obviously, some of the *neonates* (newborns) can make it without special care, but the best-case scenario involves a protected environment.

Some animals, such as the tough-as-nails llamas, are known as *precocious,* which means they're pretty much able to hit the ground running. In just a few minutes after birth, the newborn can get up and run with the herd. Other animals are *semi-precocious* in that they can stay outside in some of the elements with their mothers and are just fine. Newborn calves, for instance, can stay outside unless the temperatures drop below 0 degrees, in which case you should bring them inside.

Get a room! Places for dates and quarantines

Sometimes you need to take an animal out of the general population for any number of reasons. Perhaps the animal is sick and you don't want the illness to spread among the entire herd. Or maybe the matchmaker in you wants a particular female to breed with a particular male, so you take the animals out of their regular pens and put them into a smaller, exclusive pen so they can get up close and personal; there, they can be away from their peers, and you have some control over what's going on.

These pens can be anything from a simple fenced-in area to a small room in the barn. The important idea is that the animals are secluded from the rest of the herd, away from stress, or far enough away that germs won't spread to the other animals.

Providing horse housing

Horses really need to have a place to go into to stay warm, dry, and protected from wind and rain or snow. They just don't like being bothered by weather!

Most horse owners provide stables, or horse barns. In a stable, the horses stay in *box stalls,* small partitions that house individual animals, to ward off unwanted breeding or fighting. However, a run-in shelter (one that has at least one open side that lets the animals come and go freely) can do for horses who live in the pasture.

If you keep your horses in a stall in the barn, you can more easily monitor and groom them and also keep them protected from predators and injury. But you need to be more attentive to any breaks in walls or gates and then respond immediately to the repairs, and you have to be judicious in letting the horses out for exercise. Horses need exercise to keep them from getting restless and bored and to prevent leg problems. Barn horses can also get respiratory infections due to the close, dusty quarters. Also, barns present a fire risk — if a fire breaks out, animals locked inside may not be able to escape on their own.

Letting horses live outdoors with a three-sided run-in shelter readily available means they get more physical and mental stimulation, and you don't have to worry so much about ventilation. Grass also provides most of their diet (though making sure the grass is nutritious can involve some careful management). But monitoring what's going on with them, corralling them for saddling or grooming, and generally keeping them clean are harder.

Typically, show horses aren't allowed to be out in pasture because all that running around on uneven ground can result in injuries. You may want to consider providing a small paddock area to let your horse out to kick up his heels by himself and be a horse for a little while. *Paddocks* are outdoor spaces that allow horses some exercise but don't contain enough good-quality grass as pastures do.

You also need a place to corral horses so you can encourage them to come into a pen for medical reasons, for grooming, to prepare them for an outing, or for any other reason you may want to round them up. A round pen area is a popular design because it makes for a good place to exercise, groom, or tack them; it can also serve as a paddock area.

For more info on living spaces and shelter for horses, check out *Horses For Dummies,* by Audrey Pavia and Janice Posnikoff (Wiley).

Storing Food and Water

The animals need a place to get in out of the cold and rain, and so does the stuff you feed them. In this section, I discuss storing hay, grain, and water.

Hay storage facilities

If you have animals who don't depend on good-quality pasture grass, you likely need some sort of place to store the hay you feed them. Any of the barn designs I discuss earlier in the "Raising Barns" section can work.

Keep the hay in a different area from where you keep the animals. If you have the hay in the same areas as the animals, they may get into the hay and eat too much, as well as trample the bales, making a mess that encourages vermin and hay rot.

The most important part of a hay storage structure is the roof, which keeps rain and snow from rotting the hay. These storage areas can be open on all four walls, but you really should have a roof, which can be as simple as a tarp. You may want to put the hay on pallet (or something else that keeps the hay off the ground), providing a layer between the ground and the hay for good drainage.

Another consideration in hay storage is keeping wildlife out of it. Wildlife can mess up your hay bales, and if you don't scoop up the loose hay, it turns to rot — not a financially good way to deal with hay. At the very least, cover it with a tarp.

Grain bins

As a hobby farmer, you probably don't need a multi-thousand-gallon bin to store grain. But grain should be a part of your animals' diets, so you do need a place to store it.

Grain bins come in various sizes and shapes, and they can be built from a variety of materials. The style and size you go with depends on the number and size of animals you have. On our small farm, we buy a couple of large bags of grain and store them in large aluminum garbage cans (with lids), though your needs may vary.

Water tanks

Animals should always have access to clean water. Keeping water available in the winter when all it wants to do is freeze up is an especially arduous task. But in the heat of the summer, having an abundant supply is equally important.

You can get really high tech with your watering systems, or you can go really low tech. Fancy automatic waterers are great, but they may be too costly for a smaller operation. A clean stream on the property may be ideal, but not everyone is so blessed. A simple large tub — even an old bathtub! — is an efficient and inexpensive way to go. (See Chapter 11 for wintertime considerations for your watering systems.)

You also need to be diligent about keeping the watering bowls clean. After a few days, you get a buildup of hay (from wind or even from your tossing hay over the fence) and maybe even feces. The water containers need to be turned over and washed out regularly.

Storing Equipment

You need a place to store all those small tools that you need on the farm. (For more info on some of those gadgets you can't live without, see Chapter 6.) Here are some possible structures:

✔ **Tool sheds:** Gardening tools, animal care tools, and so on need to be organized (so you can find them when you need them) and locked up (so they don't walk away at night). You can build your own shed out of plywood or go all out with a design such as Tuff Shed (tuffshed.com). After the main structure is built, fill it with shelves, drawers, and pegboards.

✔ **Tack rooms:** Horses come with lots of gadgets, called *tack,* for grooming and riding. You need a tack room in your barn (or plenty of space in the garage or house) to store and organize items such as bridles, bits, saddles, brushes, and combs.

✔ **Machine sheds:** Larger machinery such as the backhoe or tractor also benefit from some sort of shelter. Besides a barn or large shed, you can build something like a carport or steel structure with a roof to keep your expensive toys covered.

Buy or build something that can hold up to the weather conditions on your farm. If you live in a high-snow or high-wind area, some designs may not work for you. We tried using a contraption that had a steel frame and was covered with a heavy tarp. Great idea, but we live at the mouth of a canyon where there's a lot of wind. After the first big windstorm, the poles were slightly bent and the tarp started shredding. After a few months, the frame was bent and the tarp was a pile of shredded plastic. This shelter was history.

Sheltering Plants in Greenhouses

Greenhouses allow you to extend your growing season — you can grow longer into the fall and even winter, and you can start earlier in the spring. In an area where summer growing seasons are short, you can start the seedlings so they're well on their way to maturity by the time the weather is conducive to planting them in the ground. Or you can bring bare-root trees into the greenhouse for the winter. (See Chapter 14 for more on growing in greenhouses.)

No matter how big you build your greenhouse, as you start using it, you'll likely find you need it to be bigger. You may use it for everything. Besides plants, you can use a greenhouse for processing honey in the fall or for storing sheared fleeces or processed wool. It's a good general-purpose building for keeping things out of the elements and nice and toasty warm. (Until the sun goes down; then it may get nice-and-not-so-toasty cold if not heated.)

You can find greenhouse kits or hire a contractor to build a greenhouse for you. All greenhouses should allow you to control the temperature, collect maximum sunlight, provide ventilation, and allow some sort of water supply. You can heat greenhouses with solar power or a heat pump, or you can even build a large compost pile inside that serves as a heat source. Get some great ideas on greenhouses at www.wvu.edu/~Agexten/Hortcult/greenhou/building.htm.

Going Underground with Cellars

Digging into the ground isn't only for planting crops. Besides root cellars for storing food, if you live in areas where tornados are prevalent (such as America's Tornado Alley), you'll likely want to have an underground area to run to in the case of a tornado.

Staying sheltered from the storm

Remember the classic movie *The Wizard of Oz,* where the family ran into the underground shelter and secured the door, saving them from the intensely harsh winds that ripped up the houses and other structures in the area? If you live in or are looking to move to a tornado-prone area, a good storm shelter is a must. In-ground shelters, such as a basement or storm cellar, offer the best protection.

You can visit www.fema.gov for some advice on locating or building a safe room. To find out whether you should be concerned about funnel clouds in your area, check with your intended state's resources or even on the Internet.

Keeping cool with fruit or root cellars

Before the age of electricity and refrigerators, the only way to store vegetables through the winter was in a root cellar. Today, these underground areas can be a wonderful storage area for the fruits of your labors.

The cellar environment is one that stays chilly and humid, keeping your fruits and veggies fresh and soft in both the winter and summer. The temperature and its stability depend on how deep the cellar is. At 10 feet, if you can swing it, you can maintain a constant temperature. A dirt floor helps maintain humidity.

An underground design is easiest to keep cool, though if you can't manage to build it that way, the cellar doesn't actually have to be underground. Here are some popular styles:

- ✔ **Hillside:** A room dug into a hillside, kind of like a cave with a door

- ✔ **Basement-style:** An underground room, with a shed built over it and a trapdoor allowing entrance into the cellar

- ✔ **Aboveground:** Sides packed with rocks, packed with dirt, and then covered with sod to mimic the underground atmosphere

You can visit waltonfeed.com/old/cellars.html for a lot of good information on how to build one. Or get a good book on root cellars, such as *Root Cellaring: Natural Cold Storage of Fruits & Vegetables,* by Mike and Nancy Bubel. For information on root cellar use, check out Chapter 15.

Chapter 6

Gearing Up with the Right Tools, Clothes, and Equipment

• •

In This Chapter

▶ Equipment selection tips

▶ Tools, glorious tools

▶ Machinery, vehicles, and other newfangled contraptions

▶ Clothing and shoes

▶ Safety equipment

• •

*O*wning a nice chunk of acreage and maybe a few animals means you get to play with a whole new set of grown-up toys. You've probably already used the basic hand gardening tools, but you can also get some really cool larger things (such as a backhoe or power sprayer) that don't make as much sense when you're living in the city. Some of these tools you've probably never heard of, such as a hoof knife or a crook. But the tools I discuss in this chapter can all help make life on your farm much easier.

Selecting the Right Stuff

Before you run out to the local hardware or farm store to load up on cool tools, you need to make some decisions about what you want in terms of quality, price, and the like. I know it's way cool to own some of the big items you need on the farm, but maybe it'd be more financially feasible to rent them. And if you do decide to buy, do you really *need* the top of the line? This section explains how to make sure your shopping experience is a good one.

Buying or renting

For smaller tools, buying them and adding them to your collection as you need them is probably best, especially if they're tools you're likely to use every day. For larger items, renting is definitely an option that can save you financially and in terms of storage and maintenance.

Look to the future as you're considering tools. Tools you suspect you'll use only once are good candidates for renting. Or you can buy a tool to use for a specific project and then sell it when you're done with it.

Sometimes you need to use equipment only a couple of times a year. Think about whether the tool has uses other than what you originally plan to purchase it for. The warmer months are your biggest season for using a tractor, but maybe you can get the snowplow accessory, which makes the machine a year-round tool.

Maybe you'd like to eventually buy a tractor or a backhoe but can't afford it right now, but you still have some tasks you need to take care of, so renting is a good idea. Perhaps you can rent your neighbor's tractor or go to the local equipment rental place and try out several different models (a great way to find out which features you like and which you really won't use).

Choosing quality

"You get what you pay for" applies to all aspects of life, and that means on the farm, too. You may be more apt to do the research and make an informed decision when looking at big ticket items, and you may be tempted to buy the lowest-priced option with smaller hand tools.

Although scrimping in some areas may be tempting, you'll probably find it's not a good idea in the long run. Even with smaller tools, you're better off investing in something a little better, a practice that can go miles in helping your budget. Tools made with quality forged steel last longer than the cheaper ones. Hammers made with not-so-great steel break, and garden shears of the same caliber get nicks in the blade, meaning you lose the couple of bucks you saved when you have to buy a new tool.

Aim for respected brands for the types of tools you want. Stanley, Black & Decker, Skil, and DeWalt are just a few brands known for quality for hand and power tools. John Deere, Bobcat, and Case are known names for big ticket farm implements.

Getting advice and trying items before you buy

Few things in life are more frustrating than plunking down big bucks for something that falls apart just a few weeks later. Or maybe you discover after using it that it just doesn't do what you want it to. That's why you should do some research before you buy and try handling the tools in person, even if you plan to order them online.

If buying smaller items, you can get most of what you need at the local home repair stores. If you're not tool savvy, your local hardware store or farm supply store may be the best bet. You can browse through the various tools and ask a knowledgeable clerk about which tools are best for the types of uses you have in mind.

Getting a bigger item such as a tractor or backhoe means going to a special retailer, such as a farm-equipment leasing place that sells used equipment, or even searching the classifieds. You may want to find ways to try out the machinery first. Maybe you can borrow a tractor of the model or brand you want from a neighbor. Or try renting from an equipment rental place. Or maybe the dealer will let you take a test drive. At the very least, try to talk to people who've owned and used what you're considering.

Loading Up on Basic Tools You Can Hold

Most of the tools you own and use regularly are those you can hold in and use with your hands. Everybody, whether a farmer or not, needs a hammer, screwdriver, power drill, and so on. But as a farmer, you have the opportunity to acquire several more tools. Read on for some info on handheld tools.

Building a handyman kit for all-purpose maintenance

During your days on the farm, machinery is going to break down, or you'll undoubtedly need to erect something, whether it's a major undertaking like a barn or a small job like a lean-to. Here are the components of a basic farmer's tool kit.

A portable toolbox and tool belt

It's nice to have everything you need for common tasks all in the same place, so look for a portable toolbox. So many styles and prices are available, but the most economical and easy styles are the 5-gallon bucket with bucket pouch (heavy fabric with pockets that drape over the side of the bucket) or a tool belt. Some toolboxes double as a stepping stool or sawhorse.

Tools to use with nails, screws, and other fasteners

Most fasteners don't do much good unless you have some tools to get them in and out, to tighten and loosen. Make sure the following equipment finds a place in your toolbox:

✔ **Hammer:** A basic claw hammer is an all-around tool for pounding things — things like nails for minor repairs or new construction or for putting up photos on the wall or for persuading that stubborn piece of wood to go into that little hole. A hammer with a steel or fiberglass handle lasts a long time and is less susceptible to having the head fly off. (The same is true for hatchets, axes, sledgehammers, and so on.) You may also want to have at least one smaller hammer for getting into tight spots or placing small nails or tacks.

✔ **Screwdrivers (both flat-blade/slotted and Phillips sets):** Things come unscrewed; it's a fact of life. And of course, there's no universal size or even shape to the screw heads, so keep a kit with several types of tips to accommodate all those different options.

A variety of screwdrivers come with interchangeable bits that are stored in the handle. A quality multi-bit screwdriver can help you deal with all the screw-driving problems you come across without lugging the entire toolbox around.

✔ **Cordless drill and bit set:** You need to drill holes when repairing or building a new animal enclosure or greenhouse. A cordless drill lets you wander away from the power outlet. Look for a spare battery pack you can keep charged so you don't have to stop in the middle of a project.

✔ **Wrenches (including socket wrenches):** A set of regular and/or adjustable wrenches should be in your tool kit for all sorts of tightening (or loosening) jobs. The crescent wrench is adjustable. Socket wrenches have a set of differently sized sockets that fit on the handle, or ratchet. Some wrench types can get in places that others can't.

If you have foreign vehicles, you may want to have a set of sockets sized in metric in addition to your set sized in fractions of inches.

✔ **Pliers:** Pliers can be the vise-grip type that you use to grab things that you can then pull or twist; the needle-nose type for working in tight, narrow places; or the lineman's pliers, which have a gripping point on the end as well as a sharp edge for cutting wires and the like.

Tools for measuring

"Measure twice, cut once" — that's the mantra for building and repair. Here are some measuring and marking tools you want to have on hand:

✔ **Tape measure:** Besides building and repairing jobs, tape measures are also handy for animal and garden work (for putting plants specific distances apart or for measuring your animals).

✔ **Pencil for marking:** Keep a sharpened pencil handy for marking off what you've measured. For more accurate placement of nails, for instance, use your tape measure to determine where the nail is supposed to go and then mark that place with a pencil. Voila! The nail goes where it's supposed to instead of sort-of-close.

- **Small combination square:** If you're going to build a pen for your critters (or a compost box or whatever), a combination square can help you get those angles perfectly at 90 degrees. This gadget also has a 45-degree angle to help you get perfect miters.

- **Level:** When you build a shelf or plant a fence post, making sure it's level is important. Like many hand tools, levels come in a variety of sizes. Most combination squares also include a leveling bubble, and they're great for small jobs. For larger jobs, a 3- to 4-foot level is ideal. You may want to check out newer laser levels as well.

Cutting and grinding tools

The following tools help you cut and otherwise make wood and metal objects fit where you want them to:

- **Handsaws:** A good tool kit should at the very least include a wood saw as well as a metal hacksaw. Bow saws are great for cutting up small trees and larger branches.

- **Power saws:** Saws come in so many varieties. You can choose among circular saws, tile and masonry saws, miter saws, and so many more. What you want to acquire depends on how handy you want to be.

- **Sawhorses:** These won't fit into your toolbox, but sawhorses are handy to have around — you can even make your own. You can also buy simple sawhorses made of metal, wood, or fiberglass. Or you can get something more sophisticated that can act as a clamp, like Black & Decker's Workmate Project Center.

- **Handheld grinder:** If something is too big and doesn't fit or if something is sticking out and you don't want it there, get your grinder and grind down the offending part. Adapters are available for either metal or wood.

En garde! Fencing tool

In the country, fencing usually has more to do with keeping animals in place and marking boundaries than swordplay. A good fencing hand tool, or *fencing pliers,* is a necessity — it hammers, cuts wire, tightens strung wire, and serves as pliers to twist the wire. It also has a claw for pulling out staples or untwisting wire.

WD-40

WD-40 is amazing and has tons of (unadvertised) uses. Sure, this stuff can unstick locks, but did you know it can also remove road tar from your car, keep flies off cows, or keeps rust from forming on saw blades? Get some more ideas here: www.wd40jobsite.com.

Screws, nails, ropes, and other fasteners

Nonanimal things need to be tied down, nailed down, or otherwise fastened down, so you need a handful of tools to do that for you:

- **Duct tape:** You may not think of duct tape as a tool, but you should have duct tape on hand whether you live on a farm or not. It's great for emergency repairs for shoes, hoses, and broken windows, among dozens of other things.

- **Screws, nails, and so on:** Screws, nails, staples, brads, nuts and bolts, and other fasteners all have purposes, so have a good, varied selection of all these items handy. You never know when you'll need one. Sort nails, screws, bolts, and nuts into jars or cans or even buy a special sorting bin. And remember to keep the lids on!

- **Wire:** Wire is indispensable for minor repairs on fencing, for binding things, or for hanging items.

- **Bailing twine:** Besides using this stuff to bind up a bale of hay, you can also use it to secure gates or as a temporary fix on fencing. And you can reuse it — after you cut it to free the hay for feeding, use the twine again and again.

- **Rope:** You can get a variety of sizes of rope made of sisal, manila, nylon, or other materials. Precut bagged rope comes in lengths such as 25, 50, or 100 feet. Most hardware, farm, and building supply stores also sell rope in whatever length you want.

- **Chains and padlocks:** Chains can pull things or allow you to secure loads or gates with a padlock. Like rope, chains come in various sizes, and you can purchase them prepackaged or in custom lengths.

- **Bungee cords:** These little stretchy items are perfect for temporarily securing loads, such as when you're taking garbage to the dump.

Store your bungee cords unstretched to maximize their life.

- **Ratchet straps:** Ratchet straps have a hook on one end and a *ratchet* (a leverage system) on the other — see Figure 6-1. Secure the hook into something, such as a tie-down on your truck bed, and then crank the ratchet to tighten the strap. Use these straps to straighten trees, pull and stretch fence, and secure loads.

- **Come-along winch:** Use these winches (see Figure 6-2) to pull things much like a ratcheting strap does but on a larger, heavier-duty scale. Pull, straighten, or position very heavy things easily.

Figure 6-1:
The lever-
like ratchet
mechanism
allows you
to tighten
the strap.

Figure 6-2:
Come-along
winches
allow for
heavy-duty
pulling and
straight-
ening.

Tools for electrical work

Even if you don't plan to splice wires and channel your inner electrician, you
should still get a circuit tester and/or voltmeter. The *circuit tester* tells you
whether there's a current going to the circuit being tested, and the *voltmeter*
tells you how many volts that circuit is getting. This is useful information in
helping determine whether the appliance plugged into the outlet is bad or if
the outlet itself is the problem.

Keep a supply of electrical tape as well. Besides its intended use, you can use electrical tape for color coding (it comes in different colors), for wrapping around outdoor extension cord connections (to keep the inevitable water out), or for securing rolls of wire.

Hot tools for metalwork and plumbing

If you're going to do some plumbing, repair sheet metal on vehicles, or piece together a metal fence, you may want to look into this metalworking equipment:

- **Teflon tape:** Used to seal pipe threads, Teflon tape is a plumber's best friend (well, one of them, anyway). This thin, white, nonadhesive tape helps the pipe threads to better seal on themselves by filling in the gaps between the threads. It also acts as a lubricant for the pipe threads when tightening the fittings together. Use this stuff with your PVC pipe fittings when you're installing your sprinkler system.

- **Pipe wrench:** A pipe wrench is designed to handle soft round objects (such as copper pipes) and is a must for dealing with plumbing issues.

- **Propane torch, flux, and solder:** A torch can help you repair and fuse copper pipes. You use *flux* to clean the metal joint. *Solder,* the substance that actually does the fusing, comes in the form of a metal (typically lead and tin) that melts when heated; you apply the molten metal to whatever needs to be fused. You can also use a torch (carefully) to heat up a frozen pipe that hasn't burst.

- **Small welder:** A welder fuses metal together and is thus a great tool to have for repairing or building metal items. Fence posts, hardware around doors, gates, and garages all can benefit from the use of a welder. Certain models can also double as a generator.

Using a welder requires some training, a welding helmet with a darkened visor, protective clothing, and special care. The brightness of a welding flame can cause eye damage. Welding also throws off sparks and bits of metal that can cause injury. If you don't know what you're doing, hire a professional!

Surveying Plant and Animal Care Tools

Regardless of your enterprise, you need equipment specifically to help you work with the plants and animals you decide to raise. In this section, I go over some of the smaller equipment you need to help you care for things that grow.

Essential multipurpose tools for the yard and barn

Although some of the tools you acquire are for a single purpose alone, others are versatile, whether you're using them for gardening, animal care, or property maintenance.

Wheelbarrows

A wheelbarrow can make the task of hauling small loads of stuff (dirt, animal feed, weeds, rocks, and so on) go quite a bit more quickly than if you had to make trips carrying just what fits in your arms. Wheelbarrows are also useful for mixing cement and hauling it to where you need it, as well as harvesting in the garden (bringing those fresh veggies back to the house) and hauling seedlings and potted plants for planting.

Rubbermaid makes a rather inexpensive plastic wheelbarrow with plastic wheels that I've used for years, hauling many, many loads of all kinds of heavy stuff. It's light enough that I can lift it by myself, and it has large wheels that go well over rough terrain. It wasn't designed for hauling rock, being plastic, yet I've hauled many tons of rock. (The trick is not to fill it all the way up.) I often use it to haul mulch or load mulch from the loader-bucket on the tractor. I also fill it up with weeds piled high, which I then unload in the tractor's bucket to haul away.

Tarps

Tarps, which you can use to drag leaf and pruning litter to your compost heap, are almost as indispensable as duct tape. You also use them for covering your hay or wood pile (or even vehicles!) to protect them from the elements or as a makeshift shelter for animals. A tarp can even work as a liner for a small, decorative artificial pond.

Garden hoses

Hoses are a necessity for watering plants and filling animal watering holes. Purchase high-quality ones with a lifetime guarantee; lesser quality risks leaks that can run your well dry. Get a wide diameter — at least ¾ of an inch. Even ⅝ inch cuts down on the quantity of water greatly.

Hoses have uses besides carrying water. I use them all the time as a cheap means of corralling animals, even using smaller sections as tools for training animals. (See the upcoming section titled "Tools for animal care" for info on corralling.) You can even use a piece of old hose to help make your buckets a little easier to carry. Slice the hose lengthwise and slip it over the metal handle of the bucket. Fasten it with duct tape, and now you have a padded handle!

Rakes

Rakes come in several styles, such as the yard rake (with a triangular head that's good for cleaning up debris and raking leaves) or the garden rake (with longer, metal tines that are good for working the dirt). You can also use rakes to clean tree branches after pruning or following a particularly bad windstorm. Rakes can help keep animal pens clean and level as well.

Sharp things

Here are some good all-around tools for digging, moving things, and cutting:

- ✔ **Shovel:** A shovel is indispensable for digging holes for planting, but it's also useful for cleaning out an animal pen, freeing your vehicle from the mud or snow, and killing rattlesnakes. Those of you who live in a snowy area need a snow shovel or two as well.

- ✔ **The mighty pitchfork:** Most of the common farm animals (see Chapter 9) eat hay, and the easiest way to get that hay into the pen is with a pitchfork. It makes pitching hay a breeze so that even those farmhands who aren't the strongest have no problem. You can also use pitchforks for nonfeeding purposes, such as cleaning stalls where you use straw for bedding.

    The different styles are better for some tasks than others. I prefer the

    - • Three-tine pitchfork for straw and hay

    - • Four-tine pitchfork for manure

    - • Five-tine fork for potato patches

- ✔ **Manure fork:** These forks are similar to pitchforks but have many close tines. Use them to pick up and transfer manure as you muck out the stalls.

- ✔ **Retractable utility knife:** These dandy little knives are perfect for cutting almost anything, whether you live on a farm or not. The blade retracts for safety so you can carry the knife in your pocket, but the blade can extend and lock at various lengths. Even better, utility knives are always sharp because extra blades are stored in the handle, and you can reverse the blades to get use from both ends. Utility knives come in a variety of sizes, but Stanley's 6-inch Classic 99 is well, the classic; it's also durable — many people use the same one for well over a decade, provided they don't lose it.

Handy tools for gardens and orchards

This section covers a few of the basic tools you need for working with plants. Depending on how big you want to go with your garden, there will be others, and several different styles of the same tool may end up in your toolbox. Check out the basic list:

✔ **Hoe:** Hoes are a big help in the garden. They can help you build watering troughs in the garden, eradicate weeds, and break up tough dirt clods. You can also use them for light grading and moving dirt around.

✔ **Pick:** For removing deep-rooted weeds or digging into very hard soil (such as soil that contains clay), the pick is a godsend. The pick can also help you plant seeds and loosen packed earth.

✔ **Lopper/pruner/trimmer:** Trees, bushes, and flowering plants need to be trimmed, pruned, or lopped to keep them healthy and strong. These tools come in a variety of sizes and styles, depending on the plant getting trimmed, but generally a pruner is smaller and is used with one hand (much like scissors), whereas the lopper is a two-handed tool for cutting off larger stems. To tackle really large branches or limbs, get a saw.

Fiskars makes relatively inexpensive pruners and loppers that are well worth the money. Felco, a Swiss company, has the top-of-the-line pruners for serious gardeners, though they're pricey.

✔ **Hedge and hand shears:** Shearing is different from trimming in that you don't just snip a bit off here and there. A shearing tool lops off growth, such as when you cut a hedge so that the edge remains at a certain height.

✔ **Pest sprayer or duster:** Pests (in both the insect and the weed form) need to be dealt with, and that means either chemical sprays or organic dusts. Styles of these pest removers range from handheld to backpack to ride-on. If you're using the same sprayer for both bugs and weeds, be sure to clean them completely.

✔ **Corn knife:** The corn knife (see Figure 6-3) is indispensible for cutting weeds and the like. I also have what we call the *bean hook,* which is a wooden handle with a sharpened metal hook on the end that quickly dispenses with weeds.

Figure 6-3:
A corn knife with a long, flat blade.

Roto-tillers are ideal for turning soil in your garden, and string trimmers can help keep your yard neat and trimmed. See the upcoming section titled "Powering Up with Bigger Tools for Bigger Jobs" for details.

For more info on garden tools, see *Gardening Basics For Dummies,* by Steven A. Frowine and the National Gardening Association (Wiley).

Tools for animal care

Keeping animals on your farm is a joy as well as a responsibility. Animals need regular care, and in many cases, that means more than just throwing some hay across the fence and keeping their water troughs full.

Taking care of hair and nails

Even animals need care for their hair and nails. Here's some of the equipment you need to keep your critters well-groomed:

✔ **Hoof trimmers:** The animals' feet need attention just like their bellies do. Like human toenails, hooves need to be trimmed. For filing down horses hooves, a rasp (file) is typically used, whereas for animals with cloven hoofs (most popular farm animals other than horses), you need clippers (see Figure 6-4).

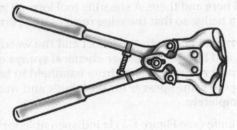

Figure 6-4:
A clipper for trimming hooves.

✔ **Hoof knife or pick:** A hoof knife or pick (see Figure 6-5) can help you deal with the stuff that gets stuck in horse's hooves. Its *j*-shaped blade enables you to dig down to clean out foreign matter from the back part of the hoof or to open an abscess. Hoof knives come in right- and left-handed models. You don't need to get the top-of-the-line model, but it's best to invest in one with good quality steel that'll last and stay sharp longer than a cheaper one.

The German company Frederick Dick offers a quality short knife with a carbon steel blade and rosewood handle; it offers you more control than models with longer blades.

Figure 6-5:
The hoof pick has a hook-like blade.

✔ **Knife/trimmer sharpeners:** Using those trimming tools wears them down, and you need to sharpen them periodically.

✔ **Shears:** Animals who produce fiber (see Chapter 9) need to be sheared regularly to keep them comfortable and so you can use the fiber. Shears come in all sizes and can be either manual (a giant pair of scissors) or electric.

✔ **Shedding brushes, cards, and combs:** These tools are also helpful in gathering fiber. Shedding brushes help you extract fiber from an animal who isn't necessarily sheared, such as an Angora rabbit. You brush the animal, and the loose fibers come out. You then use carders and combs to process the fiber into a useable form. (See Chapter 17 for more information on fiber processing.)

✔ **Grooming brushes:** Grooming brushes keep the horse's mane and tail nice and pretty.

Keeping critters in place

Many times, you need to get the critter into a corral to trim toes, give shots, prepare for a walk, and so on. Usually, animals don't go willingly. Try some of these clever items to get your four-legged friends where you want them to go:

✔ **Crooks:** A crook is a long-handled tool with a curved piece at one end. You use the curved end to grab the animal (but not horses) and persuade her to go in a certain direction.

✔ **Lariat/lasso:** If you really want to feel like a rancher, practicing using a lariat the way the real cowboys do it. This is a very effective way of getting a rope around an animal, but it takes a lot of practice.

✔ **Ropes, old hoses, or PVC pipe:** With one person at each end of a hose or pipe, you can act as a moveable barrier as you move toward a larger animal and coax him to go where you want him to go.

✔ **Fishing net:** Very helpful in capturing wayward birds, a fishing net can scoop up a chicken or duck after you've cornered her somewhere.

Of course, after you catch an animal, you usually need someplace to keep him or her so you can perform various tasks. All but those animals who've been extensively handled from birth will balk at your messing with their feet. You need some sort of holding chute (see Figure 6-6) to constrain an animal while you lift the foot and go to it.

You may also choose kennels or other small holding pens. Small kennels or holding pens can act as quarantine areas for sick animals (to keep the rest of the flock healthy) or as honeymoon suites for breeding (when you want to be sure a certain pair mate).

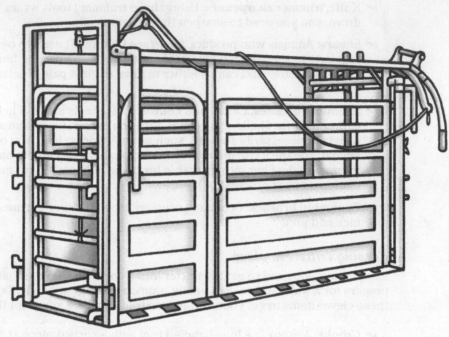

Figure 6-6:
A livestock
holding
chute.

Feeding and monitoring growth

It's helpful to keep your animals' food in aboveground feeders, usually with a drainable bottom. This keeps the feed fresh and free from the tendency to mold, lets rainwater drain out, and helps to keep vermin out of it.

A scale is useful for making sure the newborns are getting enough to eat and are growing as expected. These things can be fancy digital models or basic balancing scales (where you put the animal on a platform and load weights on the other side until the animal and weights are at the same level).

Powering Up with Bigger Tools for Bigger Jobs

Small, portable hand tools are a godsend for those small tasks, but sometimes you need something a little bigger to get the job done.

Yard and garden machines

Although you don't really need any of the following tools (the pioneers got by without them), some of these machines make getting your chores done much easier.

String trimmers/weedwhackers

Lots of acreage means lots of weeds, and in areas where you can't let the animals mow it down for you (such as in your garden), you want help. A string trimmer can be electric (which needs a power outlet nearby) or gas-powered (which gives you more roaming freedom).

Chainsaws

A mechanical version of the basic handsaw, a chainsaw makes wood cutting easy. You can find models designed for women, models that allow attachments, or models that can handle cutting down very large trees.

The model you choose, as with any other tool, depends on what you want to do with it. With an electric model, you're tethered to a plug and limited in how far you can travel from that plug, but you won't run out of power. If you're going to use it for cutting down trees, you want one with a large chain size. If you plan to break up smaller branches, you can go with a lighter-duty model (but as with all these toys, consider getting the biggest one you can afford and safely control — you may find uses for it you hadn't thought of that the smaller one won't be able to handle).

You also need a sharpening kit and a supply of oil for maintaining the saw.

Always wear safety glasses when using a chainsaw!

Roto-tillers

Every hobby farm with a garden needs a smaller roto-tiller to prepare soil for planting. Models range from a small human-propelled model to a larger gas-propelled one. Some roto-tillers are even tractor attachments. Depending on how much earth you want to till, you may even want two. Here's how the sizes measure up:

- The bigger ones for the ToolCat or other tractors can do once-yearly tillage and also prepare new garden ground.

- Smaller roto-tillers (about 20 pounds), such as Mantis or Stihl, are for use after the gardens are growing. You can maneuver them easily between rows or in beds and control them well — even a small person can lift one as needed. I recommend one with handles that bend up (such as the Stihl), which makes it more ergonomic, easier on the back to operate, and better for a tall person. Small roto-tillers often have a variety of attachments, such as an edger, though I prefer dedicated tools to the attachments.

Generators

Although small generators can't power your house in the case of a power outage, they're portable and thus are handy for helping you with power tools out in the fields. With remote power, you can light up an area, shear your animals, or use your power drills on a broken fence post.

Large generators can hold you over during power outages, which seem to happen at the most inconvenient times (such as when the outside temperatures are below 0 degrees or above 100 degrees. The farther you are from town, the more likely you'll be waiting longer for it to come back on. A gas-powered generator can save the day (as well as your comfort and the things in your freezer).

And then you can find generators that can be hooked up to a tractor. These generators rely on the tractor's power to generate electricity. Talk to store clerks and fellow farmers to see which types and models may best meet your energy requirements.

Maintenance and cleaning devices

Powered equipment can help keep your other tools sharp, clean, and running in optimal condition. Here are some to consider:

- ✔ **Small bench grinder:** A bench grinder (see Figure 6-7) gives you more control and thus more precision than a hand grinder. Use bench grinders to keep your cutting tools sharp.

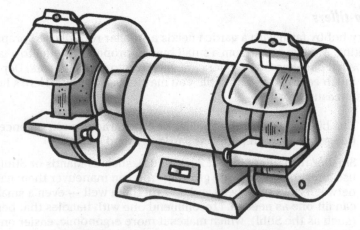

Figure 6-7:
A bench grinder.

Going online with technology tools

Getting a computer may seem to be an act that goes against your idea of moving out into the country and leaving the trappings of city life behind. But computers can really be a great help in running your hobby farm. At the very least, your computer can help you budget, write letters, keep track of inventory and sales, and do your taxes — basic office stuff. If you have an Internet connection, your computer can also help you stay in touch with the latest farm technology and be a great resource for all sorts of situations you may encounter on a given day.

Getting a bit more technology savvy may have you setting up Webcams in your animal enclosures (to see how pregnant or new moms are doing), setting up a weather station, or wiring your greenhouse so it's automatically controlled for watering, temperature, and humidity. Pretty cool stuff.

Computers on the Farm (agebb.missouri. edu/cotf) is an annual convention that may be worth a trip to Missouri (unless you already live there — then it's just a jaunt) to get the scoop on what you can computerize and what's new.

✔ **Power washer/pressure washer:** Instead of scrubbing with your hands and arms, just point the power washer at your house, barn, stall walls, tractors, and the like, and the pressure of the water takes the place of your own muscle power.

✔ **Air compressor:** An air compressor is essential not only for blowing up flat tires but also for blowing grass out of the mower's shield, cleaning dust out of tractor's air-filter, and much more.

More helpful farm tools

You'll undoubtedly come up with a few other indispensible tools that I haven't listed, but here are a couple more handy power tools:

✔ **Post-hole digger:** Digging holes for fence posts can take forever, especially if your soil has a lot of clay or rocks in it. You can get a hand digger, a gas-powered one (a two-person machine), or a tractor attachment to help you get the job done.

✔ **Snow removal equipment:** If you live in an area that gets a lot of snow, investing in some sort of snow removal equipment can save your back and a lot of time. Now that you live out in the country, you have more than just a driveway to deal with — you have animal pens to get to or much longer paths to get out to the road.

Moving Out with Tools You Can Drive or Pull

Ah, the joys of cruising your land in a tractor. The big machines and attachments in this section do the job a whole lot faster (a tractor with a mower attachment cuts grass in a fraction of the time you'd need with a push mower) or do things you simply can't do by hand (you don't need Herculean strength to move large rocks with a backhoe). And besides their power, they're just fun to ride.

Vehicles

Some jobs are just too big for the smaller tools, and you have to bring in the big guns. Whether or not you need any of these big boys depends on what you have on your property and what you intend to do with it.

A pickup truck

For hauling small amounts of hay, weeds, or other gardening debris or for pulling an animal trailer, a pickup truck is a must. Get a four-wheel drive model so you can drive through mud and snow in an emergency.

I recommend at least an 8-foot bed, not only so you can carry larger items but also for stability over rough terrain. The truck should also be equipped with a hitch so you can haul that trailer. The best-selling pickup is the Ford F-150, and I've owned many, but Chevrolet and Dodge also make excellent models.

 You can get a better product if you opt for an after-market hitch rather than having it attached at the factory.

A tractor with bucket or a backhoe

Depending on the size of your operation and the quality of your soil, a tractor or backhoe may be a necessity. The tractor is multi-talented, and with the proper attachments, it can till your land for a garden, clean out stalls (if the model has a bucket), dig post holes, remove snow, or mow the lawn. Tractors are better suited for tighter, narrower places. The backhoe can do the digging things that a tractor can do (dig big holes, move dirt around, clean out stalls, and so on) but on a much bigger scale. These machines are heavier — you can even crush pipes by driving over them.

Both machines are very expensive, and thus a hobby farmer probably won't be able to afford both. Which one you decide on depends on the lay of your land and what you want to do with it. Figure out what you want to use the machine for and then shop for a new or used one. Pay attention to the following:

- Size (tractors come in light duty or heavy duty, so you have a choice there)

- Types of attachments

- Cost

- Brand name (a good brand name such as John Deere, Case, Kubota, or Bobcat can mean more reliability)

- Age and use history (if looking at a used machine)

To get the maximum power during outages, most farmers generate electricity off of their big tractors — but there's a risk to it. You have to monitor it or risk ruining the machine, which is very, very expensive.

A lawnmower or tractor-mounted mower

Hobby farmers should have a zero-turning-radius lawnmower to easily get around the many trees, shrubs, gardens, and other obstacles on the property.

Whether you go for a mower that's an attachment to a tractor or a standalone mower depends on your circumstances. Pick your machines according to what you need them to do for you. If you have a lot of grass and are doing a lot of mowing, you may consider a standalone mower. If you have hilly areas, some standalone mowers have movable blades that conform to the hillsides.

Playing it cool with the ToolCat: Machines marketed to hobby farmers

The Bobcat Company is specifically targeting the hobby farmer with a relatively new offering, the ToolCat. A friend bought one when they first became available, and she loves it. Her husband was skeptical because he was accustomed to the large tractors that cost more than a house, but he's come around to loving it, too. Here are its features:

- It has a front arm to which you can attach many tools, such as a post-hole digger (useful for fence-making as well as planting trees) and a mower. Especially useful is the front bucket.

- It's a two-seater, and it operates like an automatic transmission car — no shifting. It

has a cab, and you can work with open windows or run the air-conditioner — her husband mows in air-conditioned comfort, whereas her AC comes from driving her riding lawnmower at high speeds to generate wind blast!

- It also has a cargo box on the back that you can raise to dump loads.

- It has a remote hydraulic outlet to operate other useful machines.

Another popular hobby farming tractor is the New Holland Boomer — a cute blue number favored by Oprah.

On the other hand, having a multipurpose machine is often more efficient than buying separate machines for each task. One tractor with the appropriate attachments can handle tasks such as mowing, snow-blowing, and tilling.

A friend favors the Grasshopper mower, which has a comfortable seat. Her mower is 9 feet long and mows a 61-inch wide swath. She also bought a snow-blower attachment and hood. You may see this machine on golf courses and in parks because it's so dependable. Others, such as the Dixon, are also available.

Four-wheelers and Gators

Four-wheelers are fun little vehicles that can be useful, too. You can use them as mechanical "cutting horses" to herd your livestock into holding pens, or they can haul a small cart loaded with hay.

The *Gator* is a four-wheeler on steroids. These vehicles come in varying sizes and can be quite versatile. A Gator is like a mini truck with a seat and a small flatbed on the back for hauling hay, grain, or wood — just about anything.

Vehicle attachments

Besides the large machinery, you want some other wheeled devices to help out in various capacities on the farm. The equipment here hitches to your other vehicles:

- ✔ **Animal trailer:** If you have animals, you'll need to transport them at times. You may need to take horses to a trailhead for a hike or have to give an animal for sale a ride to the new owner's property. Depending on the situation, taking a sick animal to the veterinarian rather than having the vet come to you may be easier. Trailers come in various sizes and with various amenities — some deluxe models even include sleeping quarters for the humans.

- ✔ **Utility trailer:** For times when you don't want to use the back of the truck for hauling or your load simply doesn't fit, a utility trailer can save the day.

Remember that an animal trailer and utility trailer are usually two different things. The animal trailer needs to have sides high enough to keep the animals in, and if it has hard solid sides and a roof, it needs to have to have holes for air circulation. A utility trailer doesn't have to have such sturdy walls and can be completely enclosed.

Getting New Duds: Tools You Wear

Okay, maybe you don't think of your clothes as tools, but on the farm, the proper clothing is a big part of the job. Manolo Blahniks and Dior suits don't cut it. Suiting up for an office job is usually more about appearance than anything else, but dressing properly for farm work is all about comfort, protection, and durability. You can keep the jeans and t-shirts, but you're probably going to have to add some new items to your wardrobe.

Dressing for the occasion

You want a coat that goes out only on the farm. Your work coat can get spit on if you get in the way of unhappy camelids, or you can get blood on you or slip and fall into the mud. You won't mind getting your work coat dirty because you know it won't be going to the opera anytime soon.

Just like coats, the clothing that covers the rest of you needs to be heavy-duty and able to withstand being mucked upon, so get some bib overalls or other heavy-duty work clothes as well. According to some, Carhartt (www.carhartt.com) is *the* source for work clothes. They're sold in farm and ranch or farm and garden stores nationwide. Their clothes are extremely well-made to take tough work, and they wash well. They're not cheap, but they're relatively inexpensive considering their long lifespan. I have a pair of heavy-duty insulated coveralls that I wouldn't be without when scooping snow in −35-degree weather.

Covering your feet, hands, and head

When you really get down to business on the farm, you need to make sure you have the right footwear, headwear, and gloves. Here's what you need for cover.

Rubber boots

Walking out into an animal pen means you have to walk through poop and mud. You want footwear that's devoted to that purpose alone so you don't track anything into the house. Choose good rubber boots that'll stay on your feet and let you safely move through the mucky stuff. You can choose boots that go over shoes or some that you just wear with socks.

Work boots

You need some sturdy footwear in the form of work boots (lovingly referred to as *clodhoppers* in some areas of the country). Hard-soled and even good steel-toed boots can save your toes if an animal accidentally steps on you or can help protect your feet if you step on a nail. The higher-top boots also offer ankle support, which is great when you're working on uneven ground or clodded dirt.

For leather work boots, you may want to give them a coat of linseed oil to help preserve the leather and make them more water resistant.

The Original Muck Boot Company (www.muckbootcompany.com) makes several models of work boots that are fantastic. They don't have a hard toe, but who cares? They're so comfortable, and they're waterproof. Use the boots in the winter and the shoes in the summer. They're great for mucking out stalls or any farm work for that matter.

A friend swears by work boots from the Red Wing Company (www.redwing shoes.com), which is based in Minnesota. Get the kind with the hooks for quick lace-up.

Gloves

For general farm chores, leather gloves are a must. For gardening, you can find gloves made of inexpensive cotton knit fabrics that are dipped in latex. Mine last me about one season of heavy gardening before they develop too many holes. The big upside is that these gloves have no seams to wear calluses or blisters on your hands. The downside is that the latex does make you sweat and can encourage bacterial growth. Wash them or just pitch them after a while if this happens.

Straw hat or pith helmet

Working outside in the hot sun can take a toll on you. One way to alleviate this problem is to wear a hat. Several styles and materials fit the bill, from the basic wide brimmed straw hat to the pith helmet style. Some hats are designed not only to block UV rays but also to repel heat and rain, allow air circulation, and stay on in high winds.

One style, the Sun Tamer, doesn't even touch your head directly — rather, it sits on your head via a comfortable elastic band, allowing the air to circulate. It's sunproof and waterproof as well as cool, and it has a chin strap in case of wind.

Henschel Hat Company (www.henschelhats.com) has inexpensive but wonderful brimmed hats with mesh above the brim that allows air to circulate. These hats also have chin straps and are lightweight, too.

The good old-fashioned sweat band

Whether you go for a cotton handkerchief tied around your forehead or a stretchy band, sweat bands are great for summer work. They keep sweat off of your face and out of your glasses.

Choosing Tools for First Aid and Safety

Accidents happen, and no farm (or home for that matter) should be caught without a good first aid kit, fire extinguishers, and some other items to help with minor emergencies. Of course, using the right safety equipment can keep injuries to a minimum. Here's what to have on hand.

Preventing accidents and injuries

Having the following items is common sense, but they can really save problems in the long run:

- **Safety glasses or goggles:** You need to protect your eyes from flying debris when you're hammering, sawing, or doing any number of other tasks. Getting something in your eyes while using a power tool can lead to an even worse catastrophe.

- **Hearing protection:** Running motors and high-pitched squeals can get pretty annoying and can actually damage your hearing. Earplugs are the simplest form of hearing protection, but you can also get noise-cancelling earphones like the ones you see on people working near airplanes.

- **Insect repellent:** Not only can protecting yourself from bug bites make you more comfortable, but it can also save your life by helping you avoid West Nile virus and other illnesses.

- **Sunscreen:** Another consequence of being outside a lot is your exposure to the sun and its damaging rays. Save your skin from a painful burn or skin cancer down the road by slathering on the sunscreen whenever you can.

Every tool, big or little, comes with potential hazards. Adding power and size to a tool increases the need for caution. If you've never used a particular tool before, ask to see how to use it before you buy. Then, at home, take it slow until you get a good feel for how the tool handles.

Handling injuries

Although you can gather the basic items and make your own first aid kit, it's usually easier to purchase a commercial, OSHA-approved first aid kit. They come with the basics (or you can go for some pretty fancy models), and they're all packed into a handy little box.

Have more than one first aid kit around the farm — one in the house, one in the barn, and so on. Most cooperative extension services suggest having one in every vehicle and building.

Also get some *hemostats* — little medical clamps used to clamp off things that are bleeding, such as a newborn's umbilical cord (see Figure 6-8). They're great for all kinds of nonmedical things, too. You can use them to hold small pieces of food when you need to hand-feed an animal (I've used them to feed crickets to orphaned ducklings) or to pick up small items, such as a tiny bolt.

Figure 6-8:
Hemostats
clamp blood
vessels.

Chapter 7

Saving, Using, and Making Power on the Farm

In This Chapter

▶ Cutting down on energy costs

▶ Using electricity and gas

▶ Making your own energy

Most U.S. farms are at least connected to the electrical power grid. Since 1936, Uncle Sam has encouraged power companies to extend lines into rural areas. However, power outages are a reality, and regardless of your situation, it's probably prudent to have a backup power system or two. Also, many farms lack connections to natural gas lines and the utilities provided in cities. You may be so far out that alternative sources (such as propane or solar power) are your only alternatives. Or renewable energy sources may even be your preference.

Besides being practical in giving you the power you need, these energy sources can save you money because you don't have to pay the power company for their services. You may even get some tax breaks for installing certain systems. And if you generate more power than you use, alternative power sources can even make you some money. In this chapter, I run through some of your energy options.

Reducing Your Energy Needs

The big push to reduce, reuse, and recycle is something you need to put to good use on a farm. You'll likely have more buildings that use power, and accessories such as automatic watering systems or equipment such as the tractor all require energy that you don't need in the city. Reducing the amount of energy you use helps you save money on power bills and reduces the strain on alternative power systems.

Many ways to save energy are pretty logical, but they may not be things you think of right off. Here are some basic energy-saving tips:

- ✔ Use sufficient insulation and make sure major air leaks in your buildings are plugged.

- ✔ Switch from incandescent to fluorescent lighting.

- ✔ When older appliances go out, replace them with more energy-efficient models.

- ✔ Keep tractors and other equipment in good working order. Replace filters, keep tires inflated, and so on.

Reducing energy needs is much more cost-effective than, say, installing an extra bank of solar panels or a second windmill. These little things can make a big difference.

Consider having an expert do an energy audit to gauge your current energy usage and suggest areas where you can make changes. Energy audits are available from some power companies, or a local government office may be able to help you locate a company that does residential energy audits.

The Powers That Be: Relying on Traditional Sources

Except in very rare instances, most farms have access to electricity. And if you're lucky, your hobby farm may also be connected to other easily accessible utilities, such as gas, water, and sewage. Although being on the grid is a great convenience, you still need to be aware of some issues regarding how to manage your usage and cost.

Of course, if you need energy in a more remote location, you may have to rely on deliveries of propane gas. In this section, I discuss using power that comes from utility companies and gas providers, and I talk a little bit about wood-burning stoves.

Electricity

Beyond Amish communities, most people consider electricity to be a necessity on a farm. But be aware that usage and costs are likely to be significantly higher than in the city. Farms have more things that need to be plugged in, and that equipment tends to use a lot more juice than a toaster or espresso machine! Water heaters in the animal pens, grain elevators, and fans for ventilation all require electricity or some other sort of power. You may feel the pinch of higher energy prices a little more now that you're in the country.

Your power may come from an electric co-op. In the 1930s, power companies started running lines to rural areas, but even then, a lot of farms were still too far away for the companies to accommodate them, so the idea of an electricity cooperative came into being. By borrowing money from the Rural Electrification Administration (REA), locals were able to have lines run out to the houses in their areas. Making lower-cost power available to these electric cooperatives has been a big help in promoting economic development and has also offset the cost of serving sparsely populated areas. These co-ops are still in operation today in some areas.

If power lines don't extend to your property, you may have to spend a chunk of change to have the power company put new lines in your area. One possibility for a new farm is to go in with a neighbor who's also just building so you can share the costs of having new lines brought out.

In the event of a power outage, you may not be able to go very long without electricity because things like food storage depend on electricity. A backup source such as a generator is a necessity out in the country.

Natural gas

Having access to natural gas can be wonderful. If you're lucky, you may have heating bills that are lower than either the electricity or the propane alternative.

However, using natural gas may not be feasible if your property isn't close enough to a service line. Even in cities, not all areas are served by natural gas lines, and extending a line to your house can cost you thousands of dollars. The price depends on how many feet away from a current line the company needs to go. In the city, there may be several feet to travel; in the country, several hundred.

Propane

Probably the most popular form of alternative fuel in rural areas is propane. If you drive along country roads and take a look in the yards of the houses you pass, you're likely to see large cylindrical tanks (see Figure 7-1). These tanks hold the fuel that powers the furnace and maybe the inside stove, fireplace, water heater, or outside barbeque grill. A tank in your backyard can hold enough propane for several months' worth of heating and cooking, eliminating the need to hook into any lines provided by natural gas companies.

Propane tanks come in various sizes — typically 250, 500, or 1,000 gallons — and you can rent them or buy them outright. The size you decide on depends on the size of your house (or whatever you want to heat). Ask for a size recommendation at the place where you get your tank.

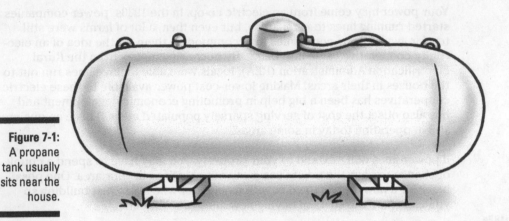

Figure 7-1:
A propane tank usually sits near the house.

Typically, the default for appliances is to use natural gas, but you can easily have an adapter installed, turning your fireplace, for instance, into one that can run on propane. Propane can also fuel emergency electrical generators.

Filling 'er up

A big truck has to come to the house regularly to fill the propane tank. You can arrange for the company to check on you regularly and fill as needed or you can call them yourself when you're getting low. We live on a dirt road that's often a mess (covered with snow or consisting of several inches of mud), so the company we use doesn't come unless we call.

If the truck doesn't come out regularly, you have to monitor the gas level yourself. After living in your house and using propane for a while, you get a feel for how much you use. But in the beginning, you should make periodic checks to see how fast the tank is emptying.

Check your propane level every couple of weeks or once a month. If the weather is good and getting the truck to your house won't be a problem, you can wait till the level is pretty low (maybe 10–15 percent). But in cold weather, when more people require fill-ups, you may want to call as soon as it gets to around 30 percent so you have a few days' buffer time to wait while you get on their schedule.

You typically pay for the propane as it's delivered, but you may be able to get around paying high wintertime prices. The company I use has a sale in August, and that's the time to fill 'er up to the brim. Even though we don't need all that propane yet, we save a lot by taking advantage of the sale.

Considering propane problems

Propane does have a few drawbacks:

✔ Although propane used to be a very inexpensive choice, its cost is going up. You may have few options in terms of companies to work with, which means price competition won't keep prices down.

✔ Propane doesn't burn as hot as natural gas, making it slightly less efficient than the natural stuff. Still, it produces enough heat to do the job.

✔ Unless you choose an underground tank, you have this big tank full of fuel sitting in your yard. (Some people get creative with their tanks, painting them to look like something other than a giant white sausage. I've seen one painted in camo colors, one turned into a giant cucumber, and one painted like an American flag.)

Hazmat comes to the farm

One day I was working in my home office, and I went downstairs for a cup of coffee. As I passed the stairs to the basement, where the furnace is, I smelled a very strong propane smell. Well, I forgot that it smells like that when the tank is empty, and all I could think was, "Oh no, there's a leak!" I called Hubby, but he was busy, so I talked to the receptionist, who told me to call 911 and get out of there.

The 911 dispatcher said to leave the house immediately, so as we were talking, I gathered up the dogs and cats and put them in one of my vehicles. I got in myself and started worrying because the car was pretty close to the house. I wanted to start the car and move it away from the house, but the dispatcher said I'd better not. So I gathered all the animals and went to a truck that was a little farther away, all the time a little panicky.

A few minutes later, a big fire truck came storming down our road. Three big burly guys all decked out in the outfits firemen wear jumped out and came over to me in the truck. I told them what was going on, and they began to investigate. They came back to tell me they did indeed

smell something and that they now needed to call in hazard materials unit. So we joked a little (they were probably trying to calm me down and get me to laugh instead of fall apart) while we waited for hazmat to arrive. Hazmat arrived shortly, and two guys (also pretty decked out) went into the house.

While they were inside, Hubby called. I told him what happened, and the first thing he said was, "Could the tank be empty?" I went over to the tank to check the level, and sure enough, it was at zero. I told the firemen, and they were gracious enough not to laugh at me. Hazmat came out of the house reporting that they didn't see anything. I told them what I found out, and everybody was relieved it was just a crazy lady panicking at nothing and not a gas leak that could've blown our house into the next county.

Moral of the story: Be sure to keep an eye on the propane level in the tank, and don't let it get empty. Oh, and if you do smell propane, don't hesitate to call in the troops — the firemen and hazmat team were really cool about coming out, even though it was a false alarm.

Recognizing empty tanks

Propane is naturally a gas, but houses use the liquid form, or *LPG* (liquefied petroleum gas). It sits in the tank, slowly vaporizing. The vapor is then released into the pipes and goes into your house.

Because propane itself has no odor, a small amount of mercaptan or ethanethiol is added so you can tell if you have a leak (around 1 gallon of stinky stuff to every 10,000 gallons of liquid propane). The odorant is a really nasty smelling stuff — like rotting cabbage or maybe rotting eggs — so when you smell it, there's no doubt that something's amiss with your propane supply.

As the mixture vaporizes, deposits, known as *heavy ends,* collect at the bottom of the tank. When there's no more propane to turn into vapor, the mercaptan or ethanethiol that's been sitting in the solution is sucked out and into the house. Then you get that not-so-pleasant fragrance, and you know you weren't paying attention to the fuel level and you allowed the tank to go empty.

Wood-burning and pellet stoves

Modern wood-burning stoves are a lot cleaner than those of the past. They're designed to be more efficient, requiring less firewood and producing less smoke and ash. You can choose among many sizes and styles, and they can actually be big enough to heat your whole house.

Pellet stoves are a cleaner alternative to wood-burning stoves. We just installed one in our basement. The pellets are made of dried and ground wood along with some other biomass wastes and are very efficient, giving off very little pollution. One drawback is that some models need electricity to operate.

You can also find special stoves designed to burn dry, shelled corn, which was very cheap until recently. If you have no other power, it may work for a little while. This Depression-era heat source may be cost effective, but it seems like a lot of work to me — hauling in the grain, hauling out the spent kernels, and such!

Other less-efficient options

Oil-fueled furnaces, although once popular, aren't so good of a choice these days. Besides the fact that they're more expensive to run, oil is a dirty form of fuel requiring diligent attention to cleaning and annual maintenance.

Coal burners are also still around — not so much in homes anymore, but industrial operations do use coal. Homeowners in the U.S. have moved away from using coal because it's a dirty option and because coal is a disappearing resource.

Getting some green to finance environmentally friendly options

The 2002 Farm Bill set up the Renewable Energy Systems and Energy Efficiency Improvements Program (Section 9006) to provide loan guarantees and grants to farmers and ranchers. Program funding has gone up in recent years as more people are feeling the pinch of the energy crunch.

The idea behind this bill is to provide grants or low-cost loans to help rural folks set up alternative (and sustainable) systems, thereby reducing the strain on other energy systems. It's a win-win thing. Farmers win by getting help (sometimes an outright grant) to build systems that'll save them a lot of money in the long run, and others feel the benefits of less pollution and possibly lower energy costs and fewer outages.

Besides this nationwide incentive program, each state has its own programs to support renewable energy. Check out www.dsireusa.org to find a state-by-state database of energy efficiency incentives, including loans, grants, and tax breaks.

Generating Your Own Power

Even if you have readily available electricity coming to your property, alternative off-the-grid resources are useful for helping contain costs and provide backup power if the electricity fails — and you know it's bound to fail from time to time. You may also want to add natural alternative energy options to your farm as a matter of choice in support of green initiatives. Or you may generate your own electricity to power outbuildings and equipment that's far from your house.

Most alternative sources provide primarily either heat or electricity, but some can provide both with special equipment. In this section, I introduce you to some ways to create your own environmentally friendly power.

A lot of alternative energy systems are pricey to install, but they often pay for themselves in energy savings over the years. When deciding whether to take the plunge, consider your energy needs, the size of your operation, available resources, and the potential for long-term savings. Take steps to make your home and outbuildings and farm processes energy-efficient before you install anything new, and look for loans and tax breaks.

Catching some rays with solar power

You can harness heat and light from the sun as solar power and use it to heat and light your house or other farm buildings. Or you can use it for other processes, such as generating electricity to pump water, which can then go into storage tanks for your livestock or can water far-off pastures.

People have used solar power since ancient times, as evidenced in the placement of doors and windows in primitive structures to optimize the sun's rays reaching inside those structures. The solar power of ancient times was available only when the sun was out, and there was no way to store the energy. These days, the latest technology helps you use solar power in much more inventive ways, helping you heat buildings and provide electricity to your home (though you can still definitely reap the benefits of good door and window placement, as the ancients did!).

Generating electricity

The typical solar power setup includes solar panels (see Figure 7-2), usually in a south-facing location, that gather the sunlight during the day and convert the energy into electricity. Systems that aren't hooked up to the power grid typically use batteries as a backup. When the sun isn't available — stormy or very cloudy days or nighttime — power is drawn from batteries, which are continually charged when the sun *is* available.

Figure 7-2: A solar panel is made up of *photovoltaic* (PV) cells, which change light into electricity.

This type of power system is clean and relatively maintenance-free. It gives off zero noise pollution (there are no moving parts to make noise or break down) and zero air emissions. In remote locations, solar panels are usually cheaper than having power lines run out. Still, solar power does have some drawbacks. In areas where the sun doesn't shine for long periods of time, especially in winter, this option doesn't work. And sunlight can be unreliable in changing weather conditions, so you need backup power sources, such as a battery or a connection to the electrical power grid.

Also, solar cells aren't pretty. Some communities have ordinances on what you can and can't build on your property; they want the neighborhood to maintain a semblance of aesthetically pleasing sameness. As beneficial as solar power can be, you may not be able to get a permit to erect a solar grid unit where you want one. To counteract this problem, you may want to consider options such as shingles with solar cells built in.

If you're bringing power to a location that has access to electricity from the power company, you may want an *intertie* system, which connects your system to their grid. That way, electricity from the power company serves as a backup, and you don't have to deal with batteries. If your solar system produces more electricity than you need, you can even sell that power back to your utility company!

For more information on solar power — including solar cells (photovoltaics), window and door placement (passive solar heating), and solar hot water — visit the National Renewable Energy Laboratory's site for farmers and ranchers (www.nrel.gov/learning/farmers_ranchers.html) or check out *Solar Power Your Home For Dummies,* by Rik DeGunther (Wiley).

Using the sun to heat water and dry crops

You can use solar power to heat water — no high-tech solar cells needed. These systems simply run water or another fluid through dark tubes, allowing the sun to heat the liquid, and then pump that liquid along. Some people even have plastic pipes installed in their floors so they can use that hot water to heat their home. (Find a contractor if you're interested — *radiant heat floor systems* are pretty complex.)

Using solar power with crops is a good way to get started with solar power. Try using this form of energy to both light and heat your greenhouse. Or use the sun to dry crops. Ancient peoples have used this drying method since the dawn of time, and these days, you can get solar dryers that effectively provide the same end result but reduce the amount of loss due to animal pilferage, wind and rain damage, and dust and dirt contamination. Large solar dryers are expensive, but you can make your own small crop dryers for drying fruits and veggies at home.

Getting down to earth with geothermal energy

Although *geothermal energy* — energy generated by heat stored under the Earth's surface — is an alternative energy source that you can harness and use for a farm's operations, doing so may be a little beyond what the small farmer can afford. But with new technology always coming out, geothermal energy is gaining in popularity, especially in newer houses. Initially installing the system may cost a lot, but in the long run, it can be much less expensive than more traditional fuels.

Tapping into naturally hot water

The concept of using geothermal energy is an interesting one. In some areas, the ground deep in the Earth stays hot, so the water down there can be brought up to the surface and used to heat the air that circulates in the house. The most obvious form of this natural heat source is geysers and hot springs like those found in Yellowstone National Park. Not everyone lives near a source of naturally occurring hot water like this, but if you do, you can use this hot water to help heat your house, greenhouse, or barn.

Most people don't have geysers in their backyards, but there *is* a lot of geothermal activity in the states in the Western U.S., including Alaska and Hawaii, so people living in those areas may be able to tap into it. The U.S. Department of Energy (DOE) has a program called *GeoPowering the West,* a big initiative to increase the use of geothermal power where these hot pockets are available.

Heat-pump systems: Counting on steady ground temperatures

Even if you don't have a hot water reservoir underground, you can still use geothermal energy to heat your house or barn with a heat-pump system. Most ground source heat-pump systems use pipes buried only several feet underground, where the temperature is a constant 50 to 60 degrees. The system takes the Earth's constant heat and condenses it, kind of like a reverse refrigerator.

These systems can also work for cooling. Water from below the surface, where the temperatures are much less than the summer air temperature, is pumped up to absorb the house's heat and work as a cooling agent.

Take a look at the DOE's Web site (www.eere.energy.gov/geothermal) for more information about geothermal energy.

Cow power! Using biomass/methane gas

Methane gas is produced from fermentation of food matter in an animal's stomach. You can actually collect the stuff and use it to heat and light the animals' own barns! Collecting this gas also reduces greenhouse gas emissions that are inevitable with an operation containing a lot of animals, such as dairy or swine farms.

Some farmers choose to place methane gas collectors directly on the animals, as in Figure 7-3. The costs vary greatly (both in the initial cost as well as annual maintenance), but a typical timeframe to expect some payback is seven years. However, if you're experiencing energy price fluctuation and you're environmentally conscious about your waste, then this option may be a feasible and sustainable one for you.

Figure 7-3:
A cow
wears a
methane
gas
collector.

You can also generate methane gas from manure by means of a *biomass composter,* or *biomass digester;* it's a lot like a regular composter, but the byproduct is methane gas. Because the bacteria that break down the manure live without the presence of oxygen, people often refer to this process as *anaerobic digestion.*

Biomass digesters work similarly to the way a cow or other animal digests food. Take a bunch of green matter (anything organic such as grass or yard clippings or droppings from grass-munching herbivores), remove air, and break it down to generate a gas that can burn and act as a fuel. As bacteria break down the plant matter into methane, it leaves behind slurry that's very high in nitrogen, which is a very good fertilizer. Therefore, this process produces no waste!

The downside to methane gas is that it doesn't compress into liquid, so methane gas needs a lot more storage space than liquefied gasses such as propane, which is stored in cylinders.

If you're interested in using methane gas as a power source, check the EPA's AgSTAR program (www.epa.gov/agstar) for more information.

Capturing water energy

Hydropower is a renewable energy source that you can use if you have access to running water nearby. Although you may think of hydropower in terms of the crashing waters of a very high, powerful waterfall, you really need only a little water running downhill.

Hydropower has been used in some capacity for eons. Consider those old grain mills with the big wheels that were used to grind the grain or to run sawmills. Although simple in their early forms, today's hydropower applications can get pretty technical.

Even though you can get only a small amount of wattage from your backyard stream, you can use hydropower for processes besides generating electricity. For instance, you can set up old-fashioned water wheels to grind your own grain.

However, this method of gathering energy often isn't a very practical one to depend on for a good part of your energy because the water flow may be unpredictable. Some years you have droughts, and some years you have floods.

Small-scale systems usually have minimal effects on the environment. Check out www.green-trust.org/hydro.htm for more information on micro hydropower systems.

Harnessing the wind

Wind power is generated with old-fashioned windmills or with giant modern turbines. It's a renewable, clean power source that helps reduce greenhouse gas emissions.

The only requirement you need is a steady source of wind. It doesn't have to blow at gale force, but it should be at least 4 to 5 miles per hour and should blow continually. A single windmill can produce anywhere from a couple hundred watts to several tens of thousands of watts.

Wind farms (large areas where several turbines are set up) are going up all over the country where this power is harnessed, stored, and distributed to homes and businesses in the area. You, too, can harness the power of the wind in your area and generate energy to run your house and farm. Of course, you don't have to generate electricity at all. Here are some windmill options:

✔ **Old-fashioned windmills:** You can use traditional windmills to pump water for irrigation or for stock watering. Smaller-capacity windmills are traditionally used for grinding grain or bringing water up from wells (much like the wheels powered by water). The wind causes the wheel to turn, which in turn powers a grinding wheel or provides the suction needed to extract water out of a well.

These smaller models can be as simple as a bunch of wooden blades (kind of a large wooden fan) or a lot fancier. Power-generating ones can generate a couple hundred watts when the wind is blowing.

✔ **Commercially made turbines:** If you have enough wind and enough land, you can go big time with a commercial wind turbine or two. These structures can produce a large amount of electricity for your farm, and if you're hooked into the power grid, you can sell whatever you don't use back to the local power company.

Biodiesel: Brewing your own fuel

Believe it or not, you can run a diesel engine on something other than petroleum-based diesel fuel. Alternative fuel sources have been around for a long time. Many of them were developed during World War II, with the rationing of mainstream fuel oils. Biodiesel — which is kind of a glorified, thinned vegetable oil — was one, but some engines ran on alcohol, ethanol, or even charcoal.

Biodiesel is a biodegradable, nontoxic substance that you can easily and inexpensively make yourself and use to fuel your diesel engines, such as those in the tractor, backhoe, or generator. This renewable fuel gives off less carbon dioxide than you get from petroleum-based diesels, and it can also be recycled.

A friend of a friend developed a business making biodiesel out of waste products. He gathers used oil from french fryers in the area (all those fast food places out there have tons of the stuff) and then goes through the process of turning the mixture into fuel. Because he's getting the waste oil for free, biodiesel costs him about 75 cents per gallon to make.

If you're interested in playing chemist, here are the basics on how the biodiesel-making process works:

1. **Gather your main ingredient — the waste vegetable oil — and filter out any food scraps.**

2. **Heat the oil and add methanol and lye while stirring to remove remaining impurities and take out the fatty acids.**

   If not removed, they block the fuel injectors in the engines running biodiesel.

3. **Let the mixture cool and remove the glycerin.**

   The good fuel part rises to the top, and glycerin sinks to the bottom as a gelatinous mass. Remove the mass.

4. **Wash the remaining mixture to remove any soap particles or other residue that remains.**

5. **Filter the mixture to separate the oil from the water.**

6. **Dry the remaining biodiesel.**

   Now it's ready to be used as fuel.

Check out the Collaborative Biodiesel Tutorial at `www.biodieselcommunity.org` or visit `journeytoforever.org/biodiesel_make.html` to find details on how to make your own.

One issue with biodiesel is that it *gels* (thickens, kind of like freezing) at 30 degrees, so if you want to use it when temperatures are at or below 30 degrees, you need to blend it with upwards of 80 percent of regular diesel. In temperatures over 30 degrees, you can run your engines with pure biodiesel.

Biodiesel can act as a solvent, so when you first use biodiesel, change the filters on your engines. That way, any gunk buildup that accumulated on the filter when you used regular fuel (petro-diesel) won't break down and be introduced into the biodiesel fuel.

As you brew and use biodiesel, make sure you heed the following warnings:

✔ You can store biodiesel safely, but the homebrewing process involves some caustic and flammable chemicals. Do your research before you whip up a batch, and make sure you follow safety precautions — work in a well-ventilated area, and wear heavy rubber gloves, eye protection, and a dust mask. Check to see whether local laws regulate the storage of these chemicals.

✔ Biodiesel has a tendency to degrade rubber fuel lines on tractors made before 1993, so use biodiesel on newer models.

Chapter 8

Managing and Accessing Your Land and Natural Resources

• •

In This Chapter

▶ Taking care of water

▶ Managing land access

▶ Using natural fencing and wind and fire breaks

▶ Caring for pastures

▶ Controlling erosion

• •

*N*ature can be a lifesaving force, but it can also invade and threaten your property. Having nature help you in the form of rock outcrops acting as natural barriers or of ponds for watering your animals can be great, but what about when Mother Nature gets angry and suddenly those once-helpful features become nightmares? Rivers and streams can also overflow and endanger your crops, animals, and even your house.

Of course, you're affecting nature as well. Some farming operations may require you to alter some of the more permanent features of the land. For instance, you may need to selectively remove some trees to increase your planting area or build a few roads to get to the back pastures.

In this chapter, I show you tips for managing your water resources as well as how to move and store water. I cover how to use existing features of the land as protective barriers, as well as how to build natural barriers. I also explain the importance of road maintenance on and near your property. I wrap up with discussions of grazing, pasture management, and erosion.

Getting Your Feet Wet in Water Management

Water is life, and if you're lucky enough to have a natural water source on your farm property, it can come in handy for

✔ Watering crops

✔ Providing water for your animals (for tips, see Chapter 11)

✔ Household needs, such as drinking, bathing, and laundry

✔ Suppressing fires (in the absence of fire hydrants, it provides water beyond what the fire trucks can hold)

But as with all natural resources, water needs care and management. In this section, you discover where to find water, how to transport it, and how to make sure it's safe to drink.

Considering water sources

In its most natural sources, water comes in many forms. It can roll or trickle by in a river or stream, or it can sit contentedly contained by the banks of a small pond or larger lake, or you can collect the stuff that falls from the sky. This section discusses water sources, as well as some of the challenges you may have in using them.

Rivers and streams

If you live near a well-flowing river or stream, your first thought may be something like, "Cool!" And you'll probably be right, because these natural water sources with their constantly flowing water can be a great asset on your farm. If you don't have issues with pollution from upstream sources, streams can be great for your animals — fresh water is always around, and the stream can save you a lot of work in not having to haul water.

Having a river or stream on your property is a great thing — usually. However, you may run across some problems.

Problems with Mother Nature

Rivers and streams can flood or, equally bad, they can dry up. One of my friends has a river running through his farm, and after heavy rains upriver, the water backs up on his cropland — even without rainfall on his own land. If he gets rain, too, his land can't drain if the river is already full. You have to have contingency plans to deal with these fluctuations.

Building a *levee* — an elevated flood bank with a slanted edge — can help keep river water within its banks. If you rely on a levee, be sure to keep it maintained. Smaller systems are in danger of failing and allowing the water to break through.

Clashes with Father Government

You may have quite a lot of frustration dealing with government entities about river management. For instance, last year, the river ate away about 5 acres of a friend's land. He had the means to fix the erosion problem inexpensively by putting large chunks of spare concrete on the bank, but he had delay after delay in getting approval. When the powers that be finally got around to his case, officials wanted him to spend $60,000 building terraces. The result was that he lost land worth a great deal of money — it's now part of the river, yet he still has to pay property taxes on it.

Pollution

Another potential problem with river water is that it can contain antibiotics, hormones, and other drugs that are now being used on today's livestock, as well as any chemicals or fertilizers from upstream. In addition to chemicals from agricultural operations, oil and grease that leak from vehicle engines and onto roadways, debris (from shrub clippings to the carelessly tossed out cigarette), and even muck that accumulates in snow can threaten water supplies. River water can also contain parasites.

You probably won't have a problem using streams for irrigation, but if you expect your animals to drink the water, have the water tested to see whether it's safe. Although animals can handle a lot less "nice" water than humans can, at some point not even livestock should be drinking it. If the water is deemed unsafe for animals' consumption, be diligent in erecting fences or other barriers to keep the critters from wandering over and taking a sip.

Laws and programs have sprung up to guard against blatant polluting activities, so take care that you don't pollute the stream, either.

Ponds

If you have a pond on your land, great! These old-fashioned watering holes have been lifesavers for ranchers for eons. If the water source for the pond is clean, you can use the pond for watering stock (see Chapter 11) or for drawing water for irrigation (see the later section titled "Irrigation: Getting water where you need it"). Rainwater or snow runoff naturally keeps it full, and some of the water you use to water crops seeps back into the pond.

Using temporary ponds

Depending on the amount of rain your area gets and the soil characteristics, you may end up with natural ponds in the springtime — called *vernal ponds* — from snowmelt or the runoff or heavy rains. Vernal ponds can give a controlled area a big boost of water to start out your summer watering system, especially if you have some way of corralling the water into designated areas where a flood won't cause major damage; for instance, you can line up sandbags or other barriers and direct water flow. Of course, vernal ponds can also create a mess. However, they usually dry up within a couple of weeks, so the mess is temporary.

Digging an artificial pond

If nature didn't provide you with a pond, you can make your own. The water in artificial ponds is muddy for the first year or two, and ponds need maintenance to keep them from becoming full of algae, but this type of pond is a viable source of water for animals.

Choose your site carefully. Find a flat and low area where the water can settle. The soil at the site should have at least 10 to 20 percent clay, because sandy substances don't hold water. (If your soil isn't right, you can add clay to give it better water retention.) Get up into your backhoe (or a rented one) and dig away. Make the pond as big as you want, and then fill it with water. Possible water sources include spring runoff from melting snow, underground springs, and water diverted from a nearby stream.

If you have a stream nearby and want to siphon a little off here and there to fill your pond, be sure to check with authorities first. There just may be an endangered water bug downstream that will die off if it doesn't get the right flow of water.

For an extensive discussion of pond building, see the USDA handbook "Ponds — Planning, Design, Construction" (www.nh.nrcs.usda.gov/features/Publications/ponds_planning.pdf).

Watch out for fertilizer runoff from fields, which makes ponds grow too much algae. Grasses can intercept some fertilizer before it enters the water. You may be able to stock your pond with fish to eat the algae (and mosquitoes), and then you can go fishing, too!

Natural springs

Another natural water source can come in the form of a natural spring. Springs occur when the water table isn't so far underground and water seeps or even flows up to the surface. Springs typically have clean, drinkable water — it can be the same water that comes out of a well.

Wells

If you don't have natural sources of water such as streams or ponds — and connecting to a public water system is too costly — then you need a well. The good thing about having a well is that the water is yours to use as you want to, with no monthly bill to the city and no penalties for going over a limit.

Facing drawbacks

You do face some downsides with wells. First of all, digging down to the water table can be very expensive. On my farm, the drillers didn't hit water until they'd dug a little over 200 feet. The bill is calculated depending on how far the digging goes down, and it added up to a lot more than we thought we'd have to pay.

Also, wells in an area usually connect to the same *aquifer,* or underground water table, so you may have some concerns that your well will run dry. A friend's farm has been encroached by large confinement buildings housing several thousand hogs, using unbelievable amounts of water. In my area, much of what used to be large tracts of land has been subdivided into much smaller units, all drawing off the same aquifer.

Digging a well

A reputable well-digging service can help you pick your well location, as well as check records for other wells in your area to estimate how far down they may have to dig.

After the digging is done, the well is outfitted with a pump. (Ours is electric and had to be replaced after about five years, to the tune of several hundred dollars, because the first one was faulty — and of course, out of warranty.)

Your water then has to be tested for quality. If the quality isn't up to snuff, you have to install filters. Common problems include too much iron, magnesium, or nitrate.

Consider installing a water softener in the house. *Hard water* (water that has a high mineral content) can cause buildup around your faucets, in the pipes, and even on heating elements, reducing the efficiency of pipes and appliances. Hard water also lowers the effectiveness of soaps and detergents, so you need more soap to get a lather with hard water.

Cisterns

A *cistern* is a water tank. In its simplest form, cisterns are everywhere! Toilet tanks are theoretically cisterns because they hold enough water for a flush. Hot water tanks are holding places for water — yep, that's a cistern, too. On the farm, cisterns are large water-holding tanks typically used in areas where fresh water is scarce. Underground cisterns are usually giant cement structures (sometimes as big as a small house), as shown in Figure 8-1.

Aboveground cisterns are usually made of hard plastic — see Figure 8-2.

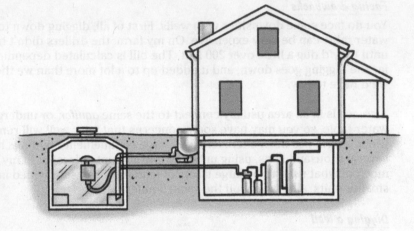

Figure 8-1:
An underground cistern.

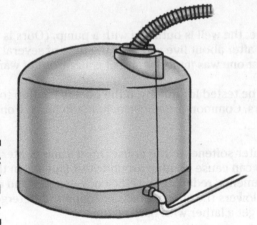

Figure 8-2:
An aboveground cistern.

Storing water for consumption

If you want to use the cistern to supply drinkable water to the house, you need to follow all sorts of regulations and receive official approval. One of the things building inspectors do is save people from themselves, so even if you intend to drink bottled water all the time, they won't give the house a certificate of occupancy and let you move in without a clean water source.

You have to keep cisterns clean and free of insects and algae. A cover can prevent dirt, leaves, animal droppings, and so on from falling into the tank. Cisterns should also be emptied and rinsed out periodically — you're going to be *drinking* this stuff, so keep it clean!

Storing water for fire suppression

My farm uses an underground cistern as a backup water source, mostly for fire suppression. Because we're not close enough to city water, in the case of a summer brush fire, we need to have a reliable source of a lot of water to make sure those fires don't eat up our house and outbuildings. In fact, installing a fire suppression tank was a requirement for getting the building permit.

Our fire suppression tank is a giant hollow concrete block, underground, that holds 5,000 gallons of water just for this purpose. It can also serve as an actual lifesaver in dire emergencies: If a fire encroaches and there's no way out, the hole is big enough for people to jump into the water and survive a firestorm passing overhead.

Supplying the cistern

You can supply cisterns from water pumped out of your well, by collecting rainwater, or by hiring somebody to deliver water to you.

Rainwater is a free, clean water source, and collecting rain is a very old idea whose time has come again. The concept is to catch runoff from your roof. The water gets corralled down through pipes (with filters to keep out the bugs and dirt) and into a lidded barrel. Rigging up special rainwater- and dew-collecting barrels can be a big help in some of your watering needs around the farm and even in the house. You can buy rain barrels at garden centers or online or have your cistern specially fitted to collect rainwater.

I have a friend in Alaska whose house is watered via a cistern that gets filled by truck as needed. Her house is very near the oceanfront and therefore lacks a freshwater table to draw from, is far enough away from the city that no city water is available, and is in an area where the sea breeze and mist carry salty deposits, so any collected rainwater isn't good for drinking. The best and only option for her is to have water delivered.

Irrigation: Getting water where it's needed

You need some means of dispersing water around the farm. You can do it the old-fashioned way by filling up a smaller vessel and hauling it yourself to where you need it, or you can use pumps and pipes with automatic devices to do the watering at a specific time of the day. In this section, I discuss irrigating your fields and garden. (*Note:* Critters need a reliable source of clean, fresh water every day. You can use any of your available natural features — a stream or pond — as long as it's free of contamination. See Chapter 11 for info on watering animals.)

Often, when people hear the term *irrigation,* they tend to think of long ditches traversing hundreds of acres in the desert or of some other large, complex method of moving water. But *irrigation* simply means artificially bringing water to the areas where you need it. Systems can range from opening a flood gate to installing a series of pipes or running small hoses through your garden.

The possibilities are endless, and what you decide on depends on how big of an area you have to water. Here are some of the more common options:

- **Flood or surface irrigation:** Some areas offer irrigation in the form of a small canal running through or on the edge of the various properties in the area. Each landowner gets a share of the water and an assigned day (or even time of the day) to open the gate and allow the water to roll over his or her land. The water soaks in and eventually dries, and then you do it all over again.

- **Sprinklers:** Sprinkler irrigation delivers smaller amounts of water. Delivery can be via a series of pipes (modern systems use PVC) with sprinkler heads. Or you can use traveling sprinklers with big wheels, a common sight in rural areas (see Figure 8-3).

  If your water is coming from a source other than your well or city water, where the possibility of debris exists, you need to filter it.

Figure 8-3:
Wheel line
irrigation
sprinklers.

- **Drip irrigation:** Instead of spraying, the water can be dripped, delivering water near the plants' roots. A hose or pipe system with a bunch of holes in it delivers water at low pressure.

- **Automatic watering systems:** Setting up an automatic watering system is a lot less work for you in the long run. It can be very simple or very complex, but the basic setup involves an underground system of pipes (usually made of PVC) that are connected to a box with valves that are controlled by an electronic mechanism usually kept in the basement, garage, or barn.

✔ **Garden hose:** The old-fashioned, human-powered hose is fine for smaller areas, but obviously, manually watering gets old if you have larger tracts of land to water. The longer the hose is, the more water pressure you lose, so watering the back forty by garden hose probably isn't very practical.

I use a combination of automatic sprinklers and the old-fashioned hose, with most of the plants watered automatically.

How often you need to water depends on where you live. If you're in the desert Southwest, you may need to give the plants a squirt every day. In places where rain is a regular sky event, you may not have to water at all. Consider getting a rain gauge to monitor how much the plants get and adjust your manual watering accordingly.

Making sure your water is okay

Before you can build on your property, an official has to test the water to make sure it's drinkable. Initially, water testing is part of the permitting process. You don't receive a permit if you're not going to be connected to the city water supply and the water on your land is unhealthy.

If your water doesn't pass the test, you have to get the required filter installed. For instance, if the water is too high in iron, you need an iron filter. In rare cases, well water doesn't pass even after filtering. If that's the case, you may be faced with digging in another area of your property or even abandoning your project.

For an existing well, the Centers for Disease Control and Prevention (CDC) recommends that you get your water tested for germs yearly and for chemicals every few years. If your house is all set and you're wondering about the natural sources for stock watering, various companies can test your well water for a fee.

Because well water comes out of the ground, you may get a slightly brownish color at times. When it's very noticeable, try not to wash whites.

Accessing All Your Land

Having dozens of acres can give you all sorts of opportunities for different operations and activities. But some of those acres may not be accessible as-is, and you'll likely need to clear some brush or even trees or at least build a road or two. This section describes some ways to access your land and how to keep those pathways in good shape.

Maintaining gravel roads and driveways

Sometimes maintaining a road that's constructed so you can get to your property becomes the responsibility of the county. In other cases, these access roads are your very own to maintain, so you may be in store for some periodic grading, snow removal, and the like.

Private access lanes

Chances are your land has some roads giving you access to the back forty, but you may have to build your own. You'll probably be able to build whatever roads you need to get around on your property without any permits or approval from the county council, but if what you do impacts someone else's access, you need to take that into consideration or even get permission.

You can improve dirt access lanes on your property by creating ditches on the sides and mounding the soil from the ditches to create a higher road, allowing rain to drain off and giving snow a better chance to blow into the ditches instead of creating impenetrable drifts on the road itself. Ditches along the sides channel water toward the natural drainage system, into rivers that ultimately flow toward the sea.

The next step is to put gravel or road base on these roads, which further enhances drainage and prevents squishy mud. Paving them is too expensive, and it's questionable how paving would hold up to the extreme conditions. Most rural folks have lanes covered with crushed limestone. This works well, but it breaks down and has to be reapplied every couple of years or more often, depending on how much heavy machinery uses it. (Just imagine farm wagons full of harvested grain and massive farm machinery using the lanes — both are very heavy.)

Gravel public roads

You may be surprised to discover that as a landowner, you actually own the side of the gravel public road that touches your property (up to the center line, which is your property line). The county gets an easement to use the road, but you, the owner, still pay taxes on the land.

You'll likely have to maintain the roads just as you do the ones on your property, though that depends on the agreement you have with the county. Heavy rains can wash out the roads or leave huge ruts in them, which require grading. I lucked out in that one of the landowners on my road works in the construction business and has access to a grader. He periodically grades the most-used portion of the road, but the ¼ mile that extends beyond his place and down to ours is up to us. Fortunately, with only a couple of vehicles going over that section regularly, it doesn't often get to a point where it's so bad that it's impassible (except for the occasional winter blizzard when huge snow drifts form!). When it does need attention, we usually just lay down some more road base using our backhoe.

Clearing land

If your land is wooded, or even if it has a lot of brush on it and you want to use the land for pasture or to build some outbuildings, you have to do some clearing. Clearing is not an easy task: You need to do some maintenance to take care that the brush doesn't grow back. You can also run into the predicament of having trees in the way that are a couple hundred years old. Bulldozing a mature forest may not sit right with you. But what are you going to do if you need to use the space?

Planning can be a big help. Before you start pulling things out, decide just how much space you need cleared. Perhaps you can work around the trees, leaving them intact and still having the space you need. Keeping some trees around is a great way to make a wind break, guard against erosion, and prevent sediment from running into any streams.

Choosing equipment

For small clearing projects, you can pull out trees and brush manually. But for larger tracts, you want to use a machine such as an excavator or a bulldozer. These usually need to be rented, though they're not cheap. Check with your local equipment rental places.

Land-clearing equipment comes in different sizes. If you have only a few trees to get out, a small bulldozer such as the Caterpillar D3 or the John Deere 450 can do. For more land, go with one of the big guns.

Disposing of the debris

If the trees are tall enough and thick enough, you may consider selling the logs to someone who can use the wood in building or furniture making. Doing that may soften the blow of taking out the trees.

If you decide to sell the logs, have the trees uprooted, knocked over, and laid on the ground for the buyer to cut and take away. If the logger comes through first, cutting off the tree and leaving a stump, excavating the stumps is a lot harder.

If you don't have anything saleable, you have to do something else to get rid of the wood you accumulate in this clearing venture. One option is to pile up the debris and burn it (but pay attention to whether your county lets you burn debris on a specific day due to air quality and fire hazards). You can then bury whatever's left over. Or maybe you can turn the branches you just pruned into mulch or take them to a yard waste recycling facility.

Recycling tree stumps

Keeping up the good practice of recycling, you can put tree stumps to use after excavation:

✔ Toss them into the bottom of your pond to create fish habitat.

✔ Cut them up into smaller pieces and use the pieces as firewood.

✔ Toss them into ditches to help with erosion control.

✔ Line them up and make a fence out of them.

✔ Turn them into a coffee table. Cut off most of the roots (leave some of the roots if they're interesting), sand the whole thing down (or keep it rough), give it a coat of stain, and then put a round or oval piece of glass on top.

Take care when burning brush or trees. Winds can kick up and start fires that can become serious. Also, fires that consume mature trees tend to be very hot and can contain hot coals that can smolder for over a week. Keep some backup water handy when you're burning and keep an eye on the coals until you're sure they're completely out (when you can put your hand into the pile and come away unscathed).

Preventing brush growth

The land should be periodically cleared with something like a bush hog, a large tractor attachment, to keep the vegetation from cropping up again (see Figure 8-4). You may even use herbicides that are targeted to certain types of foliage while allowing grass to grow.

You also want to do something to prevent erosion. See the later section "Erosion Management: Treating Your Soil Like Pay Dirt" for ideas.

Figure 8-4:
A bush hog clears brush and land.

Ensuring proper drainage

In order to use your land, you may need to remove extra water from certain areas in a way that doesn't create excessive erosion or enhance flooding. Good drainage is a must. If your soil contains too much clay, you end up with big pools after a rain. Sometimes these pools and the mud they create can interfere with your operations. Poor drainage can delay planting, thus shortening your growing season. It can cause stunted plant growth or can even kill off your whole crop. (Too much water means not enough oxygen gets to the roots.) Good drainage promotes deep root development and helps soils warm up faster in the spring.

You can take precautions to ensure proper drainage. One option is to use *drain tiles*. People used porous clay tiles back in the day, but today's drain tiles are perforated plastic pipes. The tiles are buried 2–4 feet underground at intervals. When the rain comes, the water seeps through the soil and enters small holes in the pipes; any excess water travels through these pipes to be channeled toward ditches, rivers, or ponds.

Creating and Using Natural Barriers

Natural barriers combine existing property features with features you create using the wood, stone, soil, and other materials from nature. Lots of natural barriers may be on your land, and you can use them to your advantage. They can function as natural fencing or as wind or fire breaks. This section gives you a brief survey.

Trees, shrubs, and grasses

Trees aren't just a pretty addition to your landscape. Besides providing the obvious shade, they can also aid in wind and erosion management.

Blocking sun and wind

A nice tree line is more than an aesthetically pleasing fence. Although trees don't make good fire breaks (because they can catch fire themselves), a line of trees can serve as a wind break and even protect your land and animals from fierce blowing snow. Whether you have a natural line of trees or you plant your own, these natural barriers are functional as well as beautifying elements.

Trees also provide shade to help keep your house and animals a little cooler. Especially in the hottest part of a summer day, you may notice the animals hanging out in the shade of whatever trees they can find.

Protecting streams

Sufficient vegetation along a stream can do the following:

- Intercept fertilizer-rich sediment before it reaches the stream
- Decrease erosion
- Slow flooding

Vegetative filter strips may be as simple as a line of grasses or legumes. Larger *riparian buffers* often consist of strips of trees along the banks of a stream, followed by a strip of native shrubs and a strip of native grasses. Some farmers even harvest scrap wood or plant trees that bear fruit or nuts in these buffer zones.

The government offers some incentives to plant this vegetation because riparian buffers limit pollution and provide wildlife habitat (though some programs don't let you plant orchard or Christmas trees there).

For more information on using live trees and grasses as filters, contact your cooperative extension office, the Farm Service Agency, or the Natural Resource Conservation Service.

Rocks, dirt, and hills

Large rocks and steep hillsides can possibly work as fencing material — simply design your corrals or holding pens with one of these large obstructions as a retaining wall. This doesn't work too well to keep goats in, because they love to climb, but larger animals don't typically attempt to scale a rock wall.

You can create barriers as well: Form your own hills by using a tractor or backhoe to move dirt around. Hills can be wonderful windbreaks or even fire breaks. Or stack smaller rocks and use them to make fences. Rocks can even serve as foundations for containers for anything from grain storage to animal holding pens.

On a smaller scale, *dirt berms* are mounds of dirt constructed to serve as a barrier. You can easily make them simply by piling up dirt! Berms are great for separating different parts of your garden, to separate plants that have different light, watering, and fertilizing needs.

Critter moats: Ravines, rivers, and streams

Ravine is a fancy word for a ditch or a narrow trench in the ground. If the ravine is narrow enough and deep enough, it can act as a fence "wall" that larger animals won't cross.

If wide enough, and if the water is deep and moving fast, rivers and streams can also serve as natural fencing that keeps your animals on one side. Smaller animals such as sheep and goats aren't likely to cross, and even larger animals may not be willing to venture into the water. Fast-moving water can be a big deterrent because animals' instincts tell them not to cross.

Animals can get too close to a ravine and fall in, often suffering a bone break or even death. And if a small animal decides to brave a fast-moving river, he or she is likely to be swept away. Evaluate the stability of the banks before using one of these natural barriers, and consider adding a fence not only to keep the animals from crossing and getting away but also to protect them.

Grassy Smorgasbords: Managing Pastures

A *pasture* is an area set aside for the animals to hang out and graze on whatever's edible there. You may have an area just waiting for you to put animals on it, or you may need to clear some land. (See the earlier section titled "Clearing land.")

Having pastures with grazing fodder makes the work of feeding a lot easier on you because pastures can reduce or eliminate the need to supplement the animals' diets with hay in the warmer months. But you do need to manage the pasture and keep it healthy and growing for the full season in order to get the most benefit from it.

One way to manage pastures is to rotate them — have animals graze a pasture area before you move them to the next area, allowing the plants to rest. Either use the area as foraging land one year and plant crops on it the next, or just give it a breather between grazings or plantings. Like all crops, after pastures have been picked clean, there's no more to give. The pasture land and the plants on it need a break so the foraging plants can grow back. Set up more than one pasture area. Let the animals graze on one, and then when it gets bare, move the animals to another, allowing the first one to regrow.

Allowing continuous grazing can do bad things to the land — it means the plants are continually chewed up early in life. Also, the continual trampling from so many animals can ultimately make the soil unable to produce further plant matter.

Each farm animal species has its own way of eating — and of being either kind or cruel to the pasture land. Horses, cattle, and camelids bite the grass off, whereas sheep and goats pull the grass out by the roots. Each type of livestock has different foraging needs, but a good choice of grazing mix is some sort of grass that grows well in your area with a bit of legumes.

Grass hay is a good all-around choice of pasture plant that benefits everybody. Alfalfa is okay for horses, but cud-chewing animals — such as cattle, sheep, goats, and camelids — can't have too much of it in their diets and still stay healthy. It's too high in protein, which is hard on their kidneys. Alfalfa is also higher in sugars and sweeter, so although the animals like the taste of it, too much isn't a good thing. Older horses can get diabetes from eating too sweet of a feed.

Erosion Management: Treating Your Soil Like Pay Dirt

Water, wind, or too much animal traffic can all carry away your soil, which doesn't bode well for your crops. Too much of these forces at once, and you can have a disaster. But even a little erosion at a time can reduce your layer of topsoil, make banks unstable, carry away fertilizers, and pollute rivers and streams. In this section, you find out how to keep the dirt on the farm.

In the pastures

The roots of grasses help keep soil in place, so limiting erosion in pastures depends on supporting that grass. You have to pay attention to what your land can support in terms of sheer numbers of animals. Pasture grass tolerates a limited amount of trampling, so if you try to put too many animals in a small space, the grass (and the soil underneath it) gets a workout and may get to the point where it can't recover. Although requirements vary, a good guide is to have at least 1 acre per large animal (horse or cow); you can get away with a little less for smaller animals.

Use rotational grazing to keep the soil in good shape and maintain ground cover. Let the animals graze in one pasture; then, when the forage gets low (only a couple of inches high), move them to another one to graze. Mow the forage so it's more even — kind of like cutting your hair to get rid of the weaker ends — and then let the pasture crop grow back in. When it's 6–8 inches high, it's ready for animals again.

During the winter, when nothing is growing and you're feeding animals hay, anyway, keep the critters in a separate area so they don't trample the unprotected soil.

Under the crops

You have to be concerned about erosion in your fields and gardens. Sometimes, what you do under the crops is what decides the crops' health.

Dealing with the contours of your land

One option for preventing erosion on slopes or hillsides is *contour tilling:* Instead of plowing up or down the hill, dig out narrow rows that go along the hillside at an equal elevation.

Or dig some terraces to help slow and control runoff. With *terracing,* you cut into the side of a hill with several levels or terraces so that the hillside looks like it has steps going up the side.

Forgoing the plow

Tilling soil can reduce clumps, help control weeds, and work organic matter and fertilizer into the soil, but tilling also increases erosion, and some people think it harms soil structure in the long run.

Many people are looking into *no-till farming,* in which you plant new crops over last year's residue without turning over the soil, especially for crops such as wheat and peas. By not turning the soil, mulch debris remains on the land, which helps increase moisture content in the soil. It also helps minimize soil erosion and reduces greenhouse gases because you spend less time on your tractor.

On the other hand, you need special equipment to be able to plant through the plant residue, and you have to rely more on herbicides to kill weeds. Ask your cooperative extension service or do some research on no-till farming and conservation tillage to see whether skipping the plow may be right for you.

Keeping soil in place during the off-season

Especially when the veggies aren't actively growing, a good way to protect the soil is to use cover crops. A *cover crop* is something you plant in the off season to choke out weeds and reintroduce needed nutrients into the soil, making the soil even more nutritious and ready to help crops grow big and strong. You may also plant cover crops instead of leaving a field fallow (empty) when you're using crop rotation.

In the fall, after the harvest and before the frosts (perhaps in October), plant some sort of cold-resistant crop — such as oats or annual rye — that can grow quickly in the remaining weeks of warmish weather and provide a nice cover for the soil when the really cold weather comes.

Another protective measure is leaving plant debris on the soil surface throughout the winter. Debris includes spent leaves and stems and corn stalks. After the harvest, leave this residue on the ground to help slow runoff and improve soil structure. Then remove the excess debris in the spring (till all the stuff under or burn it) and start the cycle all over again.

Around hillsides, water, and dirt roads

Hillsides and sloped banks can slowly fade away from wind that blows the dirt away, from water runoff, or from rock falls or mudslides. Also, a lot of road damage results from water runoff because the roadways are a perfect channel for rainwater. Here are some ways to protect your hills, streams, and lanes:

- ✔ The best practice is to plant a groundcover on the slopes — grass, shrubs, or even small trees (preferably native varieties) — to help stabilize the soil around waterways. For tips on the most effective planting systems, contact your extension service, look up riparian buffers and vegetative strips, or visit www.crjc.org/riparianbuffers.htm and check out their info on agricultural buffers.

- ✔ Reinforce stream banks with rocks or a thin layer of cement. The same treatment works around gravel and dirt roadways: Plant grass or lay down rocks or mulch to keep the soil in place and prevent ruts.

- ✔ Use landscape netting on slopes and hills, putting it over the dirt in hopes of keeping it intact.

- ✔ Heavy hoof prints can wreak havoc on a streambed, so try to keep animals off the banks. This is especially important during wet weather, when the soil is saturated and easily disturbed.

- ✔ Water runoff can also cause soil erosion around buildings where rain runs off the roof. Install rain gutters with downspouts and channel the runoff to areas where the dirt isn't exposed.

Part III

Calling In the Critters: What's a Farm without Animals?

The 5th Wave   By Rich Tennant

"Now that part of them is known as the gluteus maximus, or as we call it, the 'sweet spot'."

Part III

Calling in the Critters: What's a Farm without Animals?

In this part . . .

With a little acreage, you have so many opportunities for raising animals, whether you're interested in pets, food sources, help with the work, or just some good ol' country fun. This part helps you decide which animals are right for you, how to acquire them, and how to care for them.

Chapter 9

What Kinds of Critters Should You Get?

. . . .

In This Chapter

▶ Getting guard and alarm animals

▶ Catching up with your familiar four-legged or feathered friends

▶ Raising camelid fiber animals

▶ Taking a walk on the wild side

. . . .

You've probably already shared your home with a dog or cat or two. But your new surroundings open up opportunities for sharing your farm with other types of animals that you may not have dealt with in the past. A farm just isn't a farm without some sort of animals, and with all that acreage, it'd seem empty without hooves or at least some paws trotting around the grounds.

Although raising animals requires some work (a *lot* of it with some animals), work isn't the only thing farm animals bring you. Animals can enrich your life — whether they're pets, workhorses, or beings who give you something back (meat, eggs, fiber, and so on). With the proper care and attention to their needs and feelings, you and your animals can become a great team.

This chapter gives you plenty of critter options to consider. This is by no means an exhaustive list of what you can do on your farm. Take these ideas as a starting point and use your imagination to come up with something unique that works especially for you. Check with local 4-H groups, your veterinarian, and local farms for ideas on the kinds of animals that thrive in your area and get started.

Patrolling Your Borders: Animals to Help You Keep the Peace

You don't really have to have animals. Maybe you're allergic to wool, or you just don't want the extra work that having animals in your life requires. Even so, I recommend that you at least have a watchdog to help patrol your borders and a couple of barn cats to help keep the rodent population down.

Being out in open country means unwanted wildlife may visit your farm. A guard animal such as a dog can keep those larger pesky critters at bay, and a couple of cats patrolling the barn can keep it safe from the smaller pesky critters. In this section, I discuss cats, dogs, and other animals who can protect your property and alert you to unwelcome guests.

Intact animals make babies — and lots of them. Unless you want hundreds of dogs and cats running around, get them spayed or neutered.

Keeping guard with dogs

Although having a security system to protect the house is always a good idea, you can find one of the best early warning systems in a good old-fashioned man's-best-friend dog. Dogs get interested and downright excited when a strange animal, person, or car approaches your property. They even come to attention when the weather changes drastically or something else different is going on outside. Dogs are great foretellers of potential danger and are also great lifeguards for when you walk out to the barn late at night to check on something.

And as watchdogs go, size does *not* matter! A small dog can watch out and warn just as well as a big dog can. But if you're going to have larger animals, go for a bigger dog, one who's able to fend off a predator that's getting too interested in your stock.

Because of size and a few other characteristics, the Great Pyrenees is a very popular and effective choice for guarding sheep, goats, and alpacas. The breed tends to be devoted to and gentle and affectionate with its family and somewhat wary of strangers. But when not provoked, it's very mild-mannered. These dogs also tend to be nocturnal, which really helps in keeping the nighttime predators at bay.

Besides Great Pyrenees, several other breeds have been bred for the purpose of watching, including short-haired terrier breeds, Border Collies, Collies and Shelties, Labradors, and American Eskimo Dogs/Spitzes.

A watchdog (one who warns you of impending danger) and a guard dog (one who's trained to attack) are two distinct animals. Because guard dogs pose a real potential for injury to an innocent person, some states require licenses for a trained guard dog. Also, keep in mind that an innocent (or even a guilty) person who's injured by your dog can sue you for damages.

Controlling rodents with barn cats

Rodents in a barn are a nightmare if left unchecked. They can chew through wiring, insulation, and wood and can contaminate your livestock's food supply. They carry diseases that can infect any animals who spend time in your barn, and they may carry *Hantavirus,* which can infect and kill humans. So how do you keep the rodents at bay and save your investments? Enter the brave hunters living inside the small, sleek bodies of barn cats. Not all people love cats, but even those who don't can see their tremendous value in keeping the rodent population down.

When cats eat infected mice, those parasites go from the mice into the cats' bodies, who then risk getting sick (or passing sicknesses on to other animals or humans). In order to protect yourself and your protectors, do the following:

- ✔ Provide regular vaccinations and deworming, because cats are at risk for all manners of ailments. Vaccinations can prevent bite-transmitted diseases such as rabies, feline leukemia, and feline AIDS. (See Chapter 12 for information on medical attention for cats and other animals.)

- ✔ Keep their interactions with indoor pets or children minimal. Most cats who end up in the mousing job tend to be *feral* (untamed or wild) or at least outdoor cats. If you're going to get a cat specifically to be a mouser, leave him or her out in the barn. Treat these cats as working animals, not pets.

- ✔ Habitually follow any human contact with cats with hand washing, and keep the critters and unwashed hands away from your face.

A few breeds of dogs are actually very good at keeping the vermin population down. Dogs such as the Rat Terrier and Jack Russell Terrier were bred just for this purpose. However, they're also notorious for chasing other things (such as chickens), so although they can help keep the mice at bay, they can have unwanted side effects. Dogs also require daily feeding and more attention than cats. My pick would be to stick with the cats, who don't need much attention and are happy with a meal of mouse pâté every day.

Raising the alarm with peacocks, guinea fowls, and other protective animals

A few peacocks or guinea fowls can take the place of a watchdog (call them *alarm animals*) for your house because they get upset and make quite a racket when their surroundings change.

Peacocks or guineas aren't a good choice if you have close neighbors, because something as simple as a stray dog in your yard makes them go off. A stray dog can wreak havoc on your chickens, so you do want a warning, but the neighbors may not appreciate the constant updates.

In case you want some more choices, the following critters can also do a marvelous job of protecting your livestock or giving you ample warning of approaching strangers:

✔ **Llamas:** Llamas make for great watchdogs because they're big enough and tough enough to fend off small predators. Put one or two in with your smaller stock and feel assured they'll be safe. Llamas are also good at letting you know something isn't quite right outside. They stop in their tracks and stare, ears back in the direction of the strange noise or unusual animal.

✔ **Geese:** Geese can be aggressive, and they go after unfamiliar visitors who come into their territory. And because geese are rather big birds, they're capable of defending themselves and their territory against small to medium predators such as raccoons or weasels. But their biggest guarding benefit is that when they're upset, they don't stop squawking until the danger is gone. As long as you're home, you'll be notified that something is amiss out in the barnyard and that you should go check it out. Geese also have another hidden benefit that you don't get from a dog: They eat weeds and bugs.

✔ **Ducks:** Ducks can be territorial, going after and nipping at a critter who doesn't belong, but they're not so big and are no match against larger predators.

I've even heard that you can use a goat as a guard animal; however, my personal experience has shown that goats are more like prima donnas than workhorses. Our goats, anyway, are Houdinis (they *will* find the weak spot in the fence and make their getaway if at all possible), and we have to keep them in much sturdier fencing than any of our other animals.

Meeting the Common Farm Animals

Although the types of animals you can share your farm with are numerous, most people think of the ones in this section first. Here, I round up the usual suspects (but don't let this list limit your imagination — check out the later sections in this chapter for some other ideas).

For info on individual breeds of cattle, goats, horses, sheep, and the like, check out www.ansi.okstate.edu/breeds.

You may want to start out with something that won't take up a lot of your time and money while you decide whether you like raising animals. Here's how costs compare:

- ✔ Animals who require little initial output and little maintenance are the smaller ones, such as fowl (chickens, ducks, geese, and turkeys), rabbits, goats, and sheep.

- ✔ Medium- to high-cost animals are the larger ones: camelids, cattle, and pigs. Some can have a pretty high initial cost (female alpacas can cost as much as a quality horse, and a good cow can be more expensive than a poor horse), but maintenance costs are relatively low.

- ✔ The highest-cost animals in terms of initial outlay and maintenance are horses. Even though you can get horses cheaply (see Chapter 10 more on how to acquire your animals), they require a lot of maintenance.

Cattle

Cattle is a collective term referring to a group of cows, probably the number one thing people think of when they think of farm animals. (*Cow* technically refers to adult females, but calling the group of them *cows* is okay even if calves and bulls are in the group as well.)

They come in several varieties, each with its own characteristics, but generally cattle are raised for meat, milk, or help with chores. And you can't forget one more function of a cow or any other farm animal — simply having them around because you like them! Here are the major functions of cattle:

- ✔ **Dairy cattle:** Making the commitment to having dairy cattle means you have to be pretty much married to your farm (or at least you need workers who are). Dairy cattle have to be milked twice a day during the time they produce milk (usually for six to nine months after giving birth), either by hand or machine, or they go dry. The milk is then processed into milk, cheese, butter, or yogurt. Common breeds of dairy cattle include Ayrshire, Brown Swiss, Guernsey, Holstein, Jersey, and Milking Shorthorn.

Exchanging names

"What's in a name?" asks Shakespeare's Juliet. And when you start to talk about animals and all their monikers (a different one for different stages in their lives in some cases), you may be asking yourself the same thing. This table can help get some of those animals' names straightened out. *Note:* If you don't see an entry in the table, that type or stage of animal doesn't have a common specific name. For instance, a juvenile male goat is still called a *kid*.

Animal Type	Young	Juvenile Female (Unbred or Not Yet Laying Eggs)	Juvenile Male	Mature Female	Castrated Male	Intact Male
Alpacas	Cria			Female alpaca	Gelded male	Male alpaca
Cattle	Calf	Heifer		Cow	Steer	Bull
Chickens	Chick	Pullet	Cockerel	Hen	Capon	Rooster/ cock
Ducks	Duckling			Duck		Drake
Geese	Gosling			Goose		Gander
Goats	Kid	Doeling		Doe/ nanny	Wether	Buck/billy
Horses	Foal	Filly	Colt	Mare	Gelding	Stallion
Llamas	Cria			Female llama	Gelded male	Male llama
Rabbits	Kitten/kit	Doeling		Doe	Lapin	Buck
Sheep	Lamb or hogget (yearling)	Ewe lamb	Ram lamb	Ewe	Wether	Ram
Swine	Piglet	Gilt		Sow	Barrow	Boar
Turkeys	Poult			Hen	Capon	Tom

✔ **Beef cattle:** These types of cows are chosen and raised to be slaughtered. Unless you want to sell the tender meat of the calf (veal), it takes about two years before one of these guys is old enough and beefy enough to sell for food. Besides the meat you can get from your cattle (see Chapter 16), you can also tan the hide and use it for leather or suede clothing, furnishings, or even art. Common breeds of meat cattle include Angus, Herford, and Kobe.

✔ **Draft cattle:** Although the draft jobs have gone mostly to horses or machinery in modern times, a small operation can still use these beasts to help with various farming chores, such as hauling or plowing.

Another option in the cattle category is to get a mini cow. The claim is that due to their smaller size (and thus less feed consumption and less space needed per animal), beef production per acre can be greater than that from larger animals on the small farm.

Also useful is the cow's manure. (See Chapter 18 for ideas on what manure can be good for.)

Horses, ponies, donkeys, and mules

Every little kid's dream is to own a horse or a pony. Kids know nothing about what's involved in caring for them; they just love the idea of galloping through the countryside on their very own friend Flicka. For a lot of people, that dream doesn't die as they grow up. Horses, after all, can be a lot of fun. You can go backpacking with them, have them participate in rodeos or other shows, or simply entertain that childhood dream and go for a gallop through the countryside.

And don't forget their donkey and mule cousins. Donkeys, although a little smaller and probably not great for carrying adults, can be hitched up to carts. Both donkeys and mules are fun for kids' rides or for helping you around the farm.

Caring for horses is where the work part comes in, but if you love your animals, this is a labor of love. Here are a few basics. (You can check out *Horses For Dummies,* by Audrey Pavia and Janice Posnikoff [Wiley], for more-detailed help in caring for your horse.)

✔ Horses need attention and a lot of it. They need shelter from the weather; nutrient-rich food; a constant supply of clean water; clean, mucked out stalls every day; daily grooming and exercise; and good bedding (standing on a hard surface all day isn't good for them).

✔ Horses need quite a bit of room to thrive. Although horses have no set space requirements, you generally want at least 1 acre per horse.

✔ Horses are social animals who get bored and depressed or surly if they don't have regular contact with humans and other animals. If you keep them in the barn, someone needs to ride them daily.

They love treats, and bringing some into the pen or pasture means you're about to get mobbed. But don't worry — they're not carnivores, so you're safe!

Swine

A pig farm isn't something to go into lightly. Yes, you can get some very tasty meat from a pig, but the work required is probably more than is desirable for the scope of a small farm. Pork prices tend to be low (although prices vary widely, pork tends to be about 40 to 50 percent less per pound than, say, a nice steak), so small farms often have trouble recovering their investment. I read an article in which farmers were being paid to not bring the pig meat to market in hopes of reducing the supply and thus raising prices.

Here are some things to consider about raising pigs:

- ✔ If you're going for the meat angle, pigs generally need to be about 260 pounds (that takes about six months) before they have enough meat on them to butcher. That translates to a lot of food to feed and a lot of waste to clean up after!

  You can find inexpensive sources of food for pigs by going to local restaurants or grocery stores and asking for their unusable food, such as the outer leaves of lettuce.

- ✔ Baby boy pigs need to have their eyeteeth cut so they don't chew each other up. You need a veterinarian to do this. Fighting is in the pigs' nature when they get to a certain age, and if left unchecked, the fights can be deadly.

- ✔ Pigs should be in a pen, and that pen must be cleaned out regularly. (Not doing so is what leads to the stereotype that pigs stink.) Pigs have a reputation for smelling bad, and many people find pig odor particularly offensive; however, proper waste disposal and hygiene can keep odors down.

- ✔ If allowed to roam in a larger fenced-in area, they clear the land by rooting with their powerful snouts. They also scratch their backs on trees — the bark will then be history.

If you want to keep pigs on a smaller scale (one or two animals), they can be fun pets. Consider getting a Vietnamese Potbelly pig. (See www.pigs4ever.com for more information about these guys.) This breed is pretty clean (some people even keep them in the house), and individuals can have endearing personalities. A pot-bellied pig can seem like a dog sometimes as it follows you around and cuddles up to you. They love to be scratched behind the ears.

If you have guard dogs running around, make sure the pigpen is secure (so no dogs can slip under the fence) to keep both the dogs and the pigs safe from each other. Until you know how your particular animals will interact, don't let them do so unsupervised.

Sheep

Sheep are pretty easy to care for. Some people continually question their intelligence (mostly because of their strong herd mentality — where one goes, all others follow), but their reputation for feeble-mindedness isn't entirely true. They can recognize you, and each one has a personality.

Your sheep can provide you with more than wool. For instance, you can make their milk into cheeses or yogurt, and sheep's waste makes for good fertilizer that doesn't small bad the way cow manure does. Although most sheep breeds are multipurpose, some sheep are bred for specific qualities — meat, milk, or wool. For instance, Dorset sheep are good milk producers, Suffolks produce good meat, and Rambouillet sheep produce fine wool. But there are literally hundreds of breeds, and your best bet is to check what's best and what's available for the area where you live.

Getting the fiber into spinnable condition is the same process as for other fiber animals, but sheep fleeces tend to contain a lot of lanolin and thus have a more greasy feel and are a little harder to clean. Different breeds have different characteristics (like the lanolinless Icelandic or the Shetland, whose fleece is the finest of the sheep breeds), but all fiber sheep produce a warm wool that you can make into clothing or lots of other products. (Check out Chapter 17 for some ideas.)

Here are some things to consider when looking at whether you want to raise sheep:

- ✔ They eat grass. If you live in an area where grass is plentiful, your job of feeding is easy, and you don't need a lawnmower; in a dry, desert area, you have to supply all the feed.

  If you have a very large area and only a few animals, you may get by with just turning your critters loose in the pasture and letting them have at it, but it's still best to do some pasture management, the least of which is pasture rotation: Give the critters an area to graze, blocking off another area where the soil gets a break from the trampling and where you may plant a cover crop to put nutrients back into the soil.

  Grazing sheep can ruin the lawn if left unchecked because they tend to chew it down to the ground. (This behavior was one of the main causes of the sheep-versus-cattle rancher wars in the early days of the American West.) Keep feed on hand to counteract this tendency.

- ✔ Sheep must be sheared once a year, usually in mid-spring. Not shearing them causes their wool to get long and matted, making it difficult for them to get around, and you can't process matted wool at all.

- ✔ Although sheep roam freely across the countryside in some areas, it's best to fence them in (away from dogs, coyotes, and mountain lions) and put a guard animal in with them. (See the earlier "Keeping guard with dogs" section.)

✔ In the U.S., anyway, most of the meat that comes from sheep comes from the babies — the lambs — making this type of animal one of the hardest ones to let go. Lamb meat (from sheep less than a year old) is the most tender, but mutton (meat from an older sheep) can be tastier, albeit a bit tough. The sweetest lamb meat (as well as from any animal for that matter) comes from the day the lamb is weaned.

Goats

Goats are some of the most domesticated animals and have been raised for thousands of years for their meat, milk, and fiber. The babies also have pretty darn cute faces, making deciding to send one to the butcher a not-so-easy thing to do.

Meat and dairy

Depending on the condition and age of the slaughtered goat, the meat (sometimes called *chevon*) is similar to lamb, veal, or venison but is lower in fat and cholesterol. Goat meat is common in Muslim, Hispanic, Indian, Caribbean, and other ethnic cuisines.

You can drink goat's milk without pasteurization (although pasteurization is recommended). The milk also makes wonderful cheeses, yogurt, and even ice cream. The milk is lower in lactose than that of other animals, and people who are lactose intolerant can ingest it favorably.

Packers

Goats can carry stuff, so some people take them on hikes. Obviously, even small children would be too much for smaller goats, but they can carry about 25 percent of their body weight (up to 50 pounds) and can travel up to 12 miles per day, depending on terrain. They were born for the backcountry — their favorite food is brush and weeds.

The best goats to take on a hike are *wethers* (castrated males) because they don't feel the need to fight anybody and because their larger bodies (around 36 inches at the shoulder and 200 pounds) makes them good, tough packers. And they can go up to three days without water. As packers, goats are a great choice because they don't require much room to keep on your farm, and they're very affordable.

Fiber

Goats raised for their fiber have similar characteristics to meat goats in terms of their care and behavior, but their coats are vastly different. They're soft and curly, and the fiber you can get from them is wonderful. The Angora, Cashmere, and a number of crosses are the most popular breeds that fit into this category.

Despite what the Angora goat's name implies, you don't get angora fiber off its back. It actually produces a _mohair,_ which ranges from very fine and soft to coarse and scratchy (angora comes from rabbits — see the next section). Goat fleeces have little oil in them, so they're easy to clean in preparation for processing into yarn.

Cashmere is a relative term. Some say _any_ fiber goat can produce this fiber and that cashmere refers to a characteristic of the fiber itself (it must be 0.5 micrometers in diameter and be at least 3.175 centimeters long). Others say the goat that originated in the region of Kashmir is the only one whose fiber you can accurately call cashmere. People agree, however, that the fiber comes from the soft undercoat of the fiber goat. The fleece from the Kashmir-turned-cashmere is also known as _pashmina._

Care considerations

Here are some notes on goat care:

- You have to treat fiber goats as fiber goats. That means not using them as pack goats or putting them on a leash and taking them for a walk, because packs and collars cause their prized fiber to mat, and matted fiber isn't usable.

- Goats tend to be escape artists, so fences need to be strong and sturdy. Chicken wire and a couple of metal posts aren't enough — you need something stronger, such as chain link or metal fence panels.

- They can eat almost anything because their digestive systems can break down and glean nutrients from nearly any organic material. However, although they may _like_ to eat shirts off your clothesline, that doesn't keep goats healthy. They do like to eat shrubs and weeds and can even digest some plants that are toxic to cattle and horses, making them great weed clearers.

Rabbits

I'm sure you've heard the phrase _tastes like chicken_ when someone is describing an unusual meat. Well, that description holds true for the meat of the rabbit, which is low in cholesterol. Rabbits, which eat grass and leafy weeds, are more productive and are cheaper to feed than chickens. A rabbit can produce up to 1,000 percent of her body weight in food per year, which brings new meaning to the phrase _multiplying like a rabbit!_ You can skin and butcher about five rabbits in the same amount of time it'd take to do a chicken.

Rabbits also produce a fine, soft fiber. Rabbit fur is very soft, and people have used rabbit pelts for years as fur accents on coats or hats, among many other uses.

The fiber of the Angora rabbit can also be spun into a super soft yarn known simply as *angora*. Rabbits such as the Angora shed their coats a few times a year. (Yes, that means you can get the fiber and still keep the critter.) The rabbits should be brushed regularly to keep the coats free of knots, and the stuff you brush off can be added to the pile that you can later spin. For rabbits who don't shed their coats, you can shear them or you can brush or hand-pull the loose fibers out. (You know it's time to harvest the rabbit's fur when you notice clumps of wool sticking to the cage.)

Here are a few things to consider:

- Rabbits don't need much space to be happy. They should be kept confined to reduce the threat of predators.

- Take care when building their enclosures to ensure their safety. Hutches should be well above the ground and out of the way of curious dogs and other smaller critters, such as skunks.

- Because Angora hair is so long and soft, it has a tendency to mat and shed. (Remember, matted fiber is no good.) Not only does shedding mean you lose potential spinning fodder, but it can also make a mess of the cage and cling to your clothes. Regular brushing is crucial.

Small and Medium Birds: Fancying Fine, Feathered Farm Friends

The most popular type of farm poultry is the chicken, but you may want to think about others such as turkeys, ducks, geese, quails, or peacocks — these birds also produce delicious meat or eggs and can be fun to have on your farm. (On hobby farms, it's also a common practice to keep the larger members of the bird kingdom, such as ostriches and emus. See the section titled "Entering New Frontiers with Exotics and Large Game Animals," later in this chapter.)

Chickens

In meat birds, the meat builds up quickly, whereas layers put their energy into egg production. A bird who performs dual duties is heavier than an egg-layer and produces fewer eggs than a standard layer but more eggs than a meat bird. In this section, I discuss keeping chickens for eggs or meat.

Once a year, chickens *molt* (lose all their feathers), but don't worry about seeing bald chickens running around — the feathers are lost a few at a time, and it takes a couple of months to refurbish their feathery coats. They don't get bald while this is happening, but they do get rather ratty looking. Don't panic — seasonal molting is normal.

Keeping birds for eggs

Chickens kept primarily for laying eggs are called *layers*. These birds' jobs are, of course, to lay eggs, and on an average they do that about once every 25 hours — or a little less than once per day — optimally speaking. They produce less if something stresses them, such as not being fed enough or having a predator in the pen. Their bodies are smaller than the meat-producing birds and thus require less feed because they don't need to maintain muscle mass.

Some factors that affect their laying include the following:

✔ **Daylight:** If the chicken house is set up with lighting and you give them 12 or more hours of daylight, you can keep the egg production up.

✔ **Temperature:** The optimal temperature for a chicken's body to operate is between 70 and 75 degrees. As the temperature rises, the eggs produced will be smaller and have thinner shells.

Although the optimal temperature for chickens (both layers and broilers) is 70 to 75 degrees, the birds self-regulate their temperatures, and they can survive harsh conditions. We have chickens where winter temps can be below 0 degrees and summer temps may be above 100 degrees.

✔ **Molting:** In the fall, most hens go into a molt and stop laying eggs for a while.

✔ **Age:** Laying hens start laying eggs at about six months of age. Their egg production peaks at about two years of age, after which they lay fewer but larger eggs. If they get old enough, they eventually stop laying.

Note: When laying chickens get past their production years, they're often sold as stewing chickens. The meat is very tough, and stewing is the best way to make it edible.

Laying hens produce eggs regardless of whether a rooster is around because egg production is part of the female chicken's bodily functions. If eggs are your goal, you don't need for a rooster (except, of course, to give you that feeling of *really* being on a farm when the bird starts a'crowing in the wee hours of the morning). Some vegetarians are more interested in unfertilized eggs (those that are produced on farms that are rooster-free) because the eggs never have the potential of becoming a baby chicken.

Keeping chickens for meat

A chicken egg takes 21 days to hatch, and most meat chickens are bought as day-old chicks. Birds grow very fast, and the best meat is available when they're still pretty young. The idea is to get a meat chicken big as fast as possible. The longer you have to feed a chicken, the less money you make. Most meat chickens have a lifespan of six to eight weeks; thus, there's no egg production. (That doesn't happen until about six months.)

Meat birds are bred to have large breasts, so if meat birds live beyond six to eight weeks, the body gets too big and they may start to break legs or have other problems. Meat birds shouldn't be encouraged to roost, because jumping off of those roosts puts strain on their legs.

Turkeys

Turkeys, of course, are the most popular Thanksgiving meal. Turkeys need to be a bit older than chickens do at slaughter — hens (females) should be 14 to 16 weeks old, and toms (males) should be 19 to 20 weeks.

Most commercial turkeys are bred to have a lot of meat on them, and that means they're unable to fly. That also means you should provide shelter — protection not only from the elements but also from predators, because they can't fly up to the top of the barn if something gets too close.

I bet you didn't know that turkey feathers are ground up and used commercially as filler for certain animal feeds! And studies using the fibers in yarn have shown that a little bit of the fiber can make a very good insulation material.

Turkeys eat grain (with a little higher protein content that what you'd give a chicken) and have a life expectancy of about ten years.

Ducks and geese

Ducks are typically raised for their meat, eggs, and down. Their meat is a more exotic one and is often used in gourmet foods in pricey restaurants (*foie gras* is one example). Their eggs are larger and richer than that of a chicken and thus are prized among chefs.

They're a popular animal to have on the farm because they're easy to care for (they're happy with kitchen scraps), they eat bugs, and they're just fun to watch. They can also act as alarms and fend off small predators.

Geese are hardy birds and aren't as susceptible to diseases as some of their poultry cousins. They're easy to care for because they're foragers — by eating their favorite food, they help control your weeds. They even love grass clippings. Their eggs are a delicacy, and their feathers (particularly their down) make for soft and warm insulation material.

Game birds

Game birds come in several varieties, including peacocks, partridges, quail, pheasants, grouses, pigeons, doves, and guineas (see Figure 9-1). Raising them on your farm can be lucrative if you find the right markets.

Figure 9-1:
A guinea
hen.

Besides producing meat and eggs, they can also be used in hunting. People may use them for training hunting dogs or birds of prey in falconry. Or you can sell game birds to a commercial hunting operation. Such operations set the birds loose and charge hunters to come on their land and hunt them.

Some states require a license to raise game birds because they're considered protected. Check with your state to see what the rules say you can do with them.

Llamas and Alpacas: Spinning Out Info on Camelids

Camelids are popular on farms these days. The more common camelids are the llama and the alpaca, but two more critters — the guanaco and the vicuña — also fit into this family, and a few of these not-so-common family members are popping up across the U.S.

Camelids are ruminants (cud-chewers) like cattle, but they have three stomach compartments like their cousins the camels. They don't have hooves but rather walk on two toes, making them kinder to the environment when backpacking. Just looking at these guys brings joy to your spirit. They're like funny-looking horses with darling faces and curious personalities.

If you're interested in breeding these critters, you obviously need both a male and a female (although a male can be rented for a night out for your girl). But careful attention to their health and welfare can result in a lucrative money-making opportunity. A baby girl can fetch you a pretty penny — $15,000 plus. Gestation is about 11½ months.

Most people raise alpacas and llamas for their fleece. In this section, I discuss fiber animals, note the differences between the two major camelids, and talk a bit about behavior.

Raising fiber animals

No, *fiber animals* aren't made of oat bran; they're animals whose coats can be turned into yarn or felts that you can then use in a variety of applications. (See Chapter 17 for info on getting the fiber off the animals.) Depending on where in the country you live, several animals in this category can give you some very fine and fancy fleeces. The wonderful part about raising these animals is that you get to grow old with them (no need to go through the process of slaughtering) and still get something back for all those years of care and feeding.

Sicknesses can drastically reduce the quality of the fiber the animal produces. The fiber may not even be usable at the next shearing. Attention to health is of utmost importance, especially with alpacas, which are quite expensive.

A baby's first shearing produces the softest fiber. Males generally produce bigger fleeces than the females, and those fleeces take longer to become coarse. Birthing has a lot to do with fiber quality, but that doesn't mean the female fleeces aren't good; pregnant females should be shorn just prior to giving birth because this blessed event is a big stress-producer and thus a fiber-destroyer.

Stress equals not-so-good fiber, so if you just want the fiber and not reproduction, castrating the males is best because this procedure makes their coats softer due to reduced stress levels. (Competing for females and just fighting with other males is hard work!)

Harvesting hair from fiber dogs

Believe it or not, dog hair, as long as it's a long hair, can be processed and spun into yarn. Short-haired dogs such as Pit Bulls or Chihuahuas don't work, but Collies or Great Pyrenees do! The softest fur comes from the undercoat of dogs with a double coat. This type of fiber has been dubbed *Chiengora*, a combination of *chien* (the French word for *dog*) and *angora*, and it's considered a luxury fiber. Who woulda thought?

Deciding on llamas or alpacas

Which camelid is right for you? Llamas and alpacas are both good for fiber, but if you're after very high-quality fiber, go for the alpaca. If you want a dual-duty animal — not only a fiber-producer but also a beast of burden who's a lot of fun to hike with — go for the llama. Price may also come into play for you. Alpacas are more expensive than llamas, and females are more expensive than males. In this section, you discover both animal types.

The various breeds of camelids are alike enough that they can interbreed. This is frowned upon, so discourage breeding by keeping any intact males away from females.

Llamas

Llamas (see Figure 9-2) are the biggest of the camelids, and they can live as long as 29 years. They grow to about 4 feet high and up to 450 pounds, and they can carry about 80 pounds (a little less or more depending on the individual), so they're not built to carry an adult. However, they don't mind toting a small child around. And they're actually better pack animals than horses because

✔ They eat anything along the trail, so you don't have to pack much in the way of food for them.

✔ They don't spook as easily as horses.

✔ Their feet, being made of a soft material, are kinder to the environment.

They can last for as many miles as you can.

Figure 9-2:
A llama is a great pack animal.

Llamas are pretty easy to find inexpensively. (I got all six of my llamas for free, or you may find them at auctions for as little as $25.) The market for these guys just isn't what it used to be. However, some llamas bred for specific traits (such as a finer-quality fiber or very strong legs and backs) or those that are already trained (for carrying packs or following on a lead) can still bring in a decent price.

Their fiber isn't the softest around, so it's not as desirable for knitwear that goes close to the skin. However, it's very warm and is thus a great fiber for several other applications, such as outerwear (heavy coat-sweaters), horse blankets, rugs, or stuffings. You can also use it for felting or blend it with other fibers.

Alpacas

Alpacas are the queens of the camelids. Their fiber is very soft (actually one of the world's most luxurious) and is sought after in the knitwear market for next-to-the-skin garments. It comes in many colors — from pure white to several shades of browns, from grays to black — and it's greaseless, making it nearly hypoallergenic and easy to clean and spin.

Alpacas have a lifespan of about 25 years. Their cute faces and small bodies (they stand about 3 feet tall and weigh from 100 to 200 pounds) make them endearing to young and old alike.

There are two types of alpaca, each with its own fiber characteristics:

✔ **Huacaya:** The more popular Huacaya (pronounced wah-*kie*-ya) produces a sheep's wool–like fiber that's dense and soft. The Huacaya alpaca looks like a smaller, fluffier llama (see Figure 9-3).

Figure 9-3:
The fleece
of Huacaya
alpacas is
dense and
fluffy,
similar to
sheep's
wool.

✔ **Suri:** The Suri, which comprises about 20 percent of the total alpaca population, produces a longer and silkier fiber. The Suri looks like it's wearing dreadlocks (see Figure 9-4).

Figure 9-4:
Suri alpacas
have fine,
lustrous
fleece.

If you're interested only in the fiber, you can get by with just males. Educate yourself on the properties to look for, and be sure to find a reputable dealer. (Check the dealer with the Alpaca Owners and Breeders Association [www.alpacainfo.com] before making the investment.)

How much is that alpaca in the window?

Although the alpaca industry has been around since the 1980s (when the critters were first imported from South America), they're very expensive to acquire, so the market is still somewhat limited. How much dough do you need to get started in the alpaca business? Here's an idea:

✔ High-quality, proven females — those who have given birth successfully — can go for as much as $30,000 dollars.

✔ Unproven females (including new *crias* [babies]) can cost as little as $10,000 because there's no guarantee they'll be able to get pregnant or take the pregnancy to full term. You pay less and take the chance.

✔ Breeding-quality males go for $2,500 and up.

✔ Fiber-quality males (but non-breeders) typically go for $500 to $1,500.

Working around camelid temperament

A common misconception about camelids is that they spit at people. My experience is that llamas spit at *other* llamas. If you're caught in the crossfire, however, you will get hit. In this section, I explain a few other llama behavior issues.

Only a few reasons can make you the actual target. If you're trying to halter a particularly uncooperative llama, you'll likely get hit with spit. Some llamas who haven't been sufficiently halter-trained never really get comfortable with getting sheared and may show their displeasure of the process by nailing you (and this is some of the nastiest smelling stuff you ever want to experience). And if you become an implied threat to a new cria, Momma Llama may put you back in your place with a well aimed glob of llama goo. (See the nearby "My first cria and his spittin' mad momma" sidebar.)

Making friends

Camelids are herd animals, so having only one is not a good idea because they get lonely. Also, unless someone works with them from their birth (through cuddling, petting, brushing, and so on), they won't be as cuddly as they look. You may be tempted to pet them or even hug their necks. They hate that! "Just feed me and dispense with the touching" is a normal attitude.

A sure way to get the camelids to let you touch them is to approach with a handful of grain. With their faces in the grain, it doesn't matter what else is happening in the world, and you may be able to hug their necks or even kiss the tops of their heads.

Be careful when training a newborn. If you teach them that people are just big camelids and there's no need to run away, you'll end up with behavior problems down the road. With too much human interaction, males start to think that you're really just a big llama or alpaca. That's bad when they get old enough to breed because they may see you as competition and pick a fight with you if you get too close to their women, the girl camelids!

Llamas need to be trained for packing because their normal instinct is not to let you touch them. Several good books and workshops are available to help you in that arena, such as TTEAM (Tellington Touch Every Animal Method: www.camelidynamics.com). It takes a little time and love, but the results are very rewarding.

Dealing with fights

Intact (unneutered) males fight. This isn't a constant occurrence, but it's an inevitable one nonetheless. They make loud screechy noises, which can be disconcerting, especially to the neighbors. But the sound is normal, and you can't stop it short of separating them — which is a bad idea because they'll get lonely. If I were a camelid, I'd probably tell you that my opinion about my herd mate is "I hate you sometimes, but I still want you near." You may notice that very shortly after the fight, they bed down next to each other or eat side by side from the food trough.

My first cria and his spittin' mad momma

We acquired a pregnant llama from a lady who had too many animals and wanted to cull her herd. She said the llama was "probably pregnant" because the boys were "really going at her," but she had no records that would help us know when to (and whether to) expect a new arrival. So we waited.

One day I came home from a day at the local ski resort and something different caught my eye. I did a double take and saw a baby sitting on the ground. It was one of the cutest things I'd ever seen. Hubby got home shortly after I did, and we decided that it was still pretty cold at night (this was in March and a lot of snow was around), so the mother and child should get to sleep in the garage for a while.

My job was to pick up and carry the tiny, soft, sweet new baby, and it was Hubby's job to corral Momma Llama. Now, if I walked off with the baby, there was no doubt the mother was going to be right on my heels. We didn't want any surprise bolting, so Hubby haltered her. She was not happy, so while I got to snuggle (and kiss the top of the baby's head), Hubby suffered a repeated onslaught of llama spit. But both were safely ensconced in the garage (we didn't yet have a barn), and that baby is our little 3-year-old Alex today, who is the only one of the bunch who freely gives soft llama kisses.

There are exceptions to the camelids' getting-along rule. It's possible to get a rogue intact animal who, with the presence of girls in the next pen, doesn't settle down and continues to attack the other males in his enclosure. In this case, it's best to castrate this one.

Raising Things that Buzz, Squirm, and Swim

Not all farm animals are cuddly, nor do they all have feet! A few more ideas for your menagerie come in the form of some not-so-cute critters.

The critters discussed in this section all give back to their environment. They work wonderfully in conjunction with other animals and crops and thus make great supplements or side businesses for farmers who are already raising cows or goats or whatever. *Integrated agriculture* is a concept in which each piece in an operation cooperates somehow with the other pieces, and the result is a system where there's no waste — everything is reduced, reused, and recycled. On a small scale, you can use worms and fish to prepare manure for use on plants.

Honeybees: Getting the buzz on apiculture

Bees on a farm are generally bigger than those found in nature. Unlike other animals you raise for food, honeybees don't have to give up their lives in order for you to reap their benefits. Honeybees are pretty much the only insects that produce food for humans, and they play a major role in pollinating plants.

An average hive yields about 4 gallons of honey (after you leave some for the bees), though your results may be a lot more or less, depending on your particular conditions. For instance, honey production can get limited in very hot, dry climates. When temperatures go over 100 degrees, bees stop foraging and thus stop producing honey. At 100 degrees, plants are using every drop of moisture they can muster just to survive, so they don't produce nectar for the bees to collect.

Although you don't really need one of those white suits to deal with bees, you'll probably be happier if you at least have a veil to protect your face and hair. Here are some other considerations:

✔ You have to manage the hive, or the bees will die, and even careful management isn't a guarantee of success. *Beekeeping For Dummies,* by Howland Blackiston (Wiley), can give you the management details, but in a nutshell, you have to check the hives regularly and treat for diseases and pests, make sure the queen is in good health, and do routine maintenance such as cleaning and painting. Check to see that the growing hive has sufficient room for all the bees, and maybe add an additional frame. People who live in snow country have to be sure to keep the hive free of snow.

✔ A bee's lifespan is about two to three years, but that number can be cut way short for several reasons. For example, since 2006, a phenomenon called *Colony Collapse Disorder* has left many U.S. and European hives largely empty of bees — with the honey and the queen still inside. Researchers believe the culprits may be a fungus and some viruses (and not, as some scientists proposed, radiation from cellphones!).

Besides disease, changes in temperature — such as early thaws that don't stick around — can wreak havoc. When the weather gets nice and warm in February, the bees think it's time to start foraging, so they come out of the protection of the hive. But usually the cold comes back and there's no pollen to be found, so the bees die.

✔ Bees can travel up to 3 miles from their hives and are known to rob other hives of their honey. The problem with this (besides your neighbor's getting mad that his hive is empty) is that diseases can very easily spread among neighboring hives. The way to combat having your hive robbed or infected by a traveling marauder is to make sure the hive is strong and healthy. If the hive is strong, even a strong "wanna-bee" won't be able to get in and rob your bees!

✔ Even though the hive has lots of honey in it, you need to leave about 5 gallons so the bees themselves have enough nourishment to last through the winter. (They eat pollen in the summer and honey in the winter.)

A beekeeper is required by law to use standard equipment consisting of removable frames so that diseased bees can be easily taken out of the hive.

Fish: Testing the waters with aquaculture

Fish can be fun to raise because they're rather easy to care for, but what types and even whether or not you can do this depends on your state's laws. The most common types of farm-raised fish are catfish, followed by trout.

If the fish are for your own enjoyment, you have little to do in the way of caring for them. For instance, if you have a natural pond on your property, you don't have to do much besides catch them. With human-made ponds (see Chapter 8 for more on pond building), you need to take care of feeding them and keeping their homes pest free.

If you're raising fish for food, you have more health-related issues to consider, such as being careful about the herbicides you use for controlling aquatic plants and keeping the fish disease free.

Fish raised in captivity have some problems their wild counterparts do not. First of all, their diet is different, so they can have a different taste or color. They also have less room to swim and can be prone to diseases, so you have to consider using antibiotics.

Earthworms: Breaking ground with vermiculture

Worms may be the easiest of any animal on your farm. All you have to do is provide a place for them to proliferate and prosper (such as a pile of the organic matter that comes out of your other critters) and keep them moist. They turn that organic matter into worm *castings,* some of the best stuff you can use to add nutrients to the soil in your garden. Build your kid a roadside stand and sell some of the worms to local fishers. Everybody is happy!

Although worms are little digesters and can turn organic matter into nutrients, worms should not be thrown into your compost pile. A properly maintained compost pile reaches temperatures of 140 degrees, and subjecting worms to that high of a temperature means fried worms.

Instead, your worms need their own enclosure. (A simple box will do, but make sure you add air vents.) Add a bit of shredded newspaper for bedding, toss in some dirt, and add organic matter or table scraps. Put the worms on top of the pile and let them worm their way into the mix. Add more food scraps periodically so the worms always have plenty of nourishment for their own bodies and thus more stuff to make their castings out of.

Entering New Frontiers with Exotics and Large Game Animals

Besides the basic animals you usually find on a farm, you may want to consider some of the more exotic ones. Whole new worlds of benefits open up to you as you offer healthful game meats or lead people into the backcountry. There are more animals in the exotic category than I can cover in this book, and acquiring and raising them mostly depends on your environment and your state's laws on having them, but here are some of the up-and-coming names on the exotic animal circuit.

If you do decide an exotic animal is in your future, find a reputable breeder, and please be sure you're truly dedicated to having and caring for them. These exotic animals are part of the newly insurgent interest in what I call "wildlife as pets" or maybe "fun with exotic animals." Although you can find a market for their meat, eggs, fiber, feathers, and so on, you have to treat these animals with a little respect. They are, after all, relatively wild (or recently domesticated) animals. They may grow larger or stronger than you expect, leaving you with a dilemma — what do you do with the critters now?

Big birds: Ostriches and emus

Although once scarce, ostriches and emus (see Figures 9-5 and 9-6), of the *ratite* family of animals, are popping up in the U.S. more often these days, somewhat due to the old mad cow disease scare. The big birds are highly adaptable to varied climates, and their meat and eggs are prized. Although the market growth is there, these birds are still a specialty on small farms.

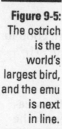

Figure 9-5: The ostrich is the world's largest bird, and the emu is next in line.

Figure 9-6: Of the two, the emu is friendlier to people.

Ostrich and emu products

Although emu and ostrich meats are gaining in popularity in the U.S., they're still a rarity, and competition with the other available meats on the market is steep. The meat is actually a red meat, and the taste resembles beef, but it's lean and contains much less fat. It's lower in calories than even chicken and turkey.

On the upside, these birds have a greater feed-to-mass ratio than cattle. *Feed-to-mass* refers to how much food they need in order to produce meat. For cows, it's 5:1 (they have to eat 5 pounds of feed to produce 1 pound of flesh), whereas with ostriches, it's 2:1 (and ostriches are happy eating bugs).

The ostrich egg can weigh 3 to 4 pounds, and a single bird lays an average of 40 eggs annually, mostly during the warm months. Emu eggs weigh a little less at 1½ pounds, and an emu produces an average of 25 eggs annually. Emus lay their eggs in the winter months.

Emu leather is exotic and is sought after in the Western boot market, among others, though ostrich hides have a more well-established market. Emu oil (fat rendered at slaughter time) is a fast-growing commodity that's being used in cosmetics and moisturizers.

Exotic feathers from ostriches, peacocks, pheasants, and the like are good for making flies for fly fishing, decorating hats, or even creating works of art.

Caring for ostriches and emus

Emus and ostriches eat leaves, roots, flowers, seeds, insects, and the occasional small lizard. In captivity, they eat whatever's available to them in the yard, but you should supplement their diets with special bird pellets, made up of vitamins and minerals, along with *browse* (non-leafy plant materials such as twigs) and vegetables such as carrots and broccoli.

These birds need a lot of room, mostly because they like to run. You also need to install tall fences because they like to jump as well.

Lean meat on the range: Buffalo and yaks

Buffalo and yak meats are considered exotic and can go for a nice price. Buffalo meat, believe it or not, has less fat than turkey or chicken! And yak meat is high in omega-3 fatty acids, which sound nasty but are actually good for your heart. Besides, these animals are just cool to have. People may bring their kids and stop in front of your farm to gaze at these majestic critters (see Figure 9-7 for an image of a yak).

The American buffalo (whose species is *Bison bison*) is often called a *bison* to distinguish it from buffaloes native to Africa.

Both buffaloes and yaks are close enough to cattle that they can mate with one another. In fact, you may run across a *beefalo* (half cow, half buffalo) or a half-yak, whose meat is less fatty than that of full-blooded cattle.

Figure 9-7:
Yaks are
gentle
creatures.

Besides meat, yaks provide fiber. The coarser outer hair is great for rugs or ropes, and the soft undercoat is comparable to cashmere. You can use the hides and fiber from yaks and buffaloes in a variety of ways, such as spinning and then knitting the fiber into a nice, warm ski sweater or using their hides for leather products.

Buffaloes are big and strong animals, so strong that high fences and corrals are in order — they can easily jump a 6-foot fence. However, if you're diligent in your feeding, including salt and mineral supplements, they're not likely to try to break out. Buffaloes, which are native to North America, have an efficient digestive system and can live on marginal pastures that would starve typical cattle. Their reproductive life is longer than cattle, and they're hardier in terms of resisting diseases.

Buffaloes and yaks can carry *brucellosis,* a disease that causes abortions and stillbirths in domestic cattle and that can also be transmitted to people in the form of undulant fever (UF). UF is a nasty thing that involves sweating and muscle pain, and it stays with you forever — you can control the symptoms, but you can never cure it completely. To prevent this, cattle need to be vaccinated when very young. (Be careful during the vaccination — if you get poked with the needle, you can get UF.)

You can find out more about raising buffaloes at the National Bison Association Web site (www.bisoncentral.com).

Venison: Domestic deer and antelope

Domestic deer and antelope require less feed and acreage than the traditional cattle herd and are pretty low-maintenance. You get a meat that's lower in fat than beef, which appeals to today's health-conscious consumers. Their antlers, shed annually, can also provide you with a little income. Deer and antelope typically have twins or even triplets, making the rate of return on these critters a little better than with cattle.

These critters are *browsers* (they like to eat woody materials such as shrubs and trees), not *grazers* (which like to eat grasses), and their feed consumption decreases in winter months. Most farmers supplement the animals' diets with pelleted food.

Deer and antelope don't really need any sort of shelter, but they do need a rather high fence because they're excellent jumpers. Fence height is regulated by state, but a typical requirement is 8 to 10 feet. Your state also has particular requirements on how much acreage you need, but a good rule of thumb is to have no more than five deer per acre. You'll most likely need a permit to raise these animals on your property.

The tendency of deer and antelope is toward the wild; thus, they need daily human contact to keep them docile.

Wooly bully: Musk oxen

Musk oxen (see Figure 9-8) are available only in very cold parts of the country because these thick-coated critters can't survive in hot climates. They're originally from Alaska, but some fiber farmers in the lower 48 (northern states only) have begun to raise them.

Their fiber is called *qiviut* and is the finest known to humankind in terms of warmth. Eight times warmer than wool and extraordinarily lightweight, a very thin layer made from qiviut is a lot warmer than the same thickness in another fiber. It comes from the undercoat of the animal and is very limited in color (a soft grey-brown). It doesn't shed and retains warmth even when wet.

Figure 9-8:
A musk ox
produces
highly prized
fiber.

Exotics just for the fun of it

A lot of exotic animals who aren't raised for meat, eggs, fiber, or feathers are happily living in homes across America, but keeping exotics as pets usually isn't a good idea. Snakes, sugar gliders, chinchillas, and even kangaroos are way up there on the cool pet factor, but in practice, these animals are usually happier living in the wild. So many times, people get all excited about a new pet, only to find that the cute baby grows into something really big, something that eats (and poops) a lot, and something that's a lot stronger than expected.

Wild animals are unpredictable and can turn on you in a flash, and they can carry all manner of diseases. You can unwittingly bring a disease into your world that causes problems for humans or other pets in the household. Wild animals commonly carry rabies or salmonella, but exotics may carry even worse diseases you don't know about. New diseases are being found all the time (remember the rodent-spread hantavirus scare?).

Also, your new pet is one fewer in the wild, one fewer who can breed and be part of a herd or other family group. Although a captive animal is regularly fed and free from predators, he or she misses out on the experience of being what that animal was born to be.

And then you have to consider the laws. Some animals have regulations on whether you can have them at all, and some require special facilities or care. For instance, as a big-cat owner, you're required to have an extra sturdy cage (preferably made of 9-gauge cyclone fencing wire), a large lockable den/box so you can safely get into the cage to clean and put out food, and double-gated entranceways. This is not only to keep the big cat in and away from your neighbors' children and livestock but also to keep a smaller animal, such a curious small dog, from unwittingly getting in (and that won't have a happy ending for the little guy).

If you must bring a wild animal into your family, the first step is to check with local laws. Then be sure to do plenty of research (on animal size, feeding requirements, lifespan, exercise needed, vet care, and so on) before deciding to spring for that bobcat or zebra.

Chapter 10

Acquiring Animals for Your Farm

In This Chapter
▶ Researching before you buy
▶ Checking credentials and looking at animals
▶ Getting an animal in the right place and the right time
▶ Finding sellers

Some of the ways for you to acquire animals provide better support than others. For instance, the current owners may help you pick just the right animal for your needs and wants or may simply be available to call afterward if you run into something you don't know what to do about. That said, if you already have some experience with a particular type of animal and you already know what to look for, maybe you just want a quick (and as-cheap-as-possible) transaction.

Some animals are very easy and very affordable to acquire; others are much more expensive and take some legwork and education so you're sure you know just what you're getting into. In this chapter, I help you decide what you want, give you advice on evaluating animals and breeders, and discuss types of sellers.

Clarifying What You Want

After you have an idea of which type of animal you want (see Chapter 9), the first thing you need to do is to ask yourself just what you want this animal to do on the farm. Are you interested in just a few companion animals? Do you want to provide meat for your family? Do you want to raise and sell or show animals? Or are you a fiber person who wants to spin yarn and knit something up from materials that came from your own animals?

Then weigh your options. For instance, say you're interested in raising animals for their fiber. Here are a few to consider:

✔ Shetland sheep, which are small and easy to handle and have a fine fiber

✔ Angora goats, whose fiber is a step up from sheep's (though the goats are a little harder to raise)

✔ Suri alpacas, whose fiber is a step up from goat's; however, the alpaca is a lot more expensive to acquire and more expensive to feed

If you're after good quality fiber, it may not matter if an individual animal isn't particularly friendly to you or if his or her legs are less than show quality. However, if you want the animal to show, then everything has to be perfect.

Some other issues that matter when you're looking for animals are age, gender, pregnancy possibilities, and what kind of shape they're in. In this section, I briefly discuss how these factors fit into your animal-buying decision.

Understanding required care and resources

Before you embark on any venture, you should get an idea of what you're getting into. Some animals require a lot of attention (horses are one), and some require extra sturdy facilities (deer need very high fences). Some need special care (for instance, fiber goats need to be shorn twice a year).

Do some research — peruse the Web, ask your neighbors, or chat with someone at the cooperative extension office to get ideas on what you can do. Find out the animal's food and shelter needs, expected lifespan, adult size, common problems, and so on.

Choosing a breed

Before you plunk down the money or the time investment in an animal, research the breeds and their common qualities. For instance, if you're interested in dairy cows, the Ayrshire is known for its easy births and strong calves and also for its longevity. The Jersey cow is a little smaller breed known for its high milk fat content and its good-tempered disposition.

After you decide on a breed, educate yourself. Look up an association (you can find tons, such as Alpaca Owners Association, the Quarterhorse Association, and so on) and see what the official breed standards are and what to expect from a certain breed. If you're looking to show and breed the animal, then you need to invest in the highest quality animal you can afford. You may not be able to resell a lower quality animal.

Considering age

The age of an animal matters. You may want to get an adult so you can see what the critter has grown into, or you may want to get a baby so you have a better chance of molding him or her behaviorally into what you want for your farm.

An older animal can have problems that may or may not be fixable. Some nutritional problems can be fixed, but that depends on how far along they are. For instance, severe malnutrition can get to a point where the animal can't recover. Or white muscle disease can be fatal in sheep and goats, but even if it isn't, these animals still have a permanent weakness. (See Chapter 12 for more information on white muscle disease.)

Babies can have special problems. If they're taken from their mothers too soon, they can have nutritional or behavioral problems; they may *imprint* on you — see you as a parent — and not know how to act as animals when it comes to breeding and the like. (See Chapter 11 for more information on behavioral considerations.) Most reputable breeders won't sell a baby till the little one is weaned (but you may have occasional opportunities to raise a really young animal if he or she is orphaned or rejected by the mother). An exception may be baby goats, who are purposely bottle-fed so they'll make friendlier pets.

Buying a trained critter may save loads of time and headaches and be a better bargain in the long run. If you've never owned a horse before, you may be tempted to acquire a young, untrained animal who has a reasonably low price tag. Hey, training the horse yourself is part of the experience, no? However, training an animal is something *you* need to be trained on, so you may have to hire someone else to train the animal and show you what to do. A beginning rider and an improperly trained horse are a dangerous combination. Sometimes an older animal is better.

Looking at gender and fertility

Keeping males, females, or both means a few differences in care. For instance, males tend to fight, especially if they're reproductively intact. If you don't care about breeding, get the males fixed. They'll be calmer, and in the case of fiber animals, you'll get better fleeces. (Stress affects the quality of fleeces — if the males are always fighting, you may see a loss of fiber quality.)

Females, of course, can get pregnant. You probably want to be there for the animals during births so you can step in if something goes wrong, even if that means being in the barn at 3 a.m. (See Chapter 12 for more information on medical care for animals.) Females may also be smaller, have a different quality of feathers or fleece, or have behavioral differences. And naturally, male birds can't exactly lay eggs.

If you want to breed the animals, you can keep both males and females (in separate pens so you can control who gets to date whom), or you can hire out your males for studs or take your girls to the neighbors for a night on the town.

Sometimes you can buy a pregnant animal and end up with a two-for-one deal. This may sound cool, but are you prepared for a newborn? Maybe you don't want infants and the extra work that goes along with them. Pregnancies can have complications, and you may lose both mother and baby.

Deciding how many you want

The number of animals you should get depends on how much land you have and how much labor you can devote to care. Whatever you decide on, ask the person you're getting an animal from or ask somebody who has animals about how many critters he or she recommends in your situation.

Some animals are herd animals and should never be isolated. It's always good to have at least two animals so they can keep each other company. And that doesn't always mean two of the same breed — you can start out with one cow in the same pen with a few sheep.

Timing Your Acquisitions and Deciding on Distance

After a long rumination, perhaps you finally decide it's time to have a cow . . . er, I mean *acquire* a cow for your farm. Before grabbing your credit card and rushing out to the 24-hour cow store, take some time to consider whether now is really the best time to get that cow, or horse, or chicken, or whatever animal you have a hankering for.

When and where you get your new animal family members is as important as what you end up with. Timing often plays a role in price and quality of animal, as well as how much work you have to do. Where you get him or her from can make a difference as well. In this section, I tell you how timing and distance can play a role in your decision.

Knowing the right time and season to buy

Sometimes, timing your acquisitions just right can save you a lot of money or trouble (or both). Consider the following time-related factors.

Season of the year

The time of the year can make a difference in the amount of work you have to do as well as in the odds that you'll raise your animals successfully. If you take on very young animals in the middle of the winter, they're tougher to raise. You have to pay special attention to making sure they're warm, especially at night.

A maybe-pregnant llama I acquired decided to have her baby girl in late winter with a lot of snow still on the ground (this was before I had the barn built). These Peruvian animals are hardy, but I didn't want to take a chance that the little girl would get too cold at night, so she and her mother spent a month in the garage with a space heater. If a garage hadn't been available, I would've had to take my chances with her being outside or have gone to some quick trouble to put up some sort of makeshift shelter.

Although my experience with a surprise llama birth worked out okay, it's much better to be prepared for the pregnant animal — have your shelter and fencing in place before you acquire the animal.

Summertime can be bad in some parts of the country, too. If temperatures get really high, you need to be able to provide shelter so the animals can get out of the scorching sun. Even a nice, big tree helps provide shade.

You can often get good prices on chicks and ducks right after Easter. Check shelters, country stores, or even pet shops. People get them for the seasonally inspired cuteness and then realize the cuteness is going to wear off.

Age of young ones

If you get a baby without his or her mother, you have even more work to do to make sure the little one is warm all day and through the night. But getting a baby means you can start training early, and you know what the baby has been through. You don't have to wonder about any bad habits an adult may have because you work with the animal from day one.

However, even if you get a trainable animal who's a little older, you can still train the critter. Horses and llamas *can* be trained later in life.

Times when farmers need to offload animals

When other people decide to go out of business or move out of town, you may be able to get some good quality animals for a relatively low cost. I got my first two alpacas this way. A friend of a friend was culling his herd due to his getting older, but he didn't want to greatly reduce the number of animals in his herd. I lucked out and got a couple of high-quality animals for a price that was on the low side of fair.

Be careful of going-out-of-business sales. Many breeders use this as an excuse to dump their poor-quality or nonproducing animals on the unsuspecting public. Don't buy anything without doing some research first.

Due to drought and then some major brush fires, the farmers in my state faced a hay shortage and thus very high prices for the hay that was available. People were getting rid of animals for very cheap because they couldn't afford (or didn't want to spend that much) to feed them. For hobby farmers who had other jobs that allowed them to pay the higher feed costs, it was a good time to get a hold of some great animals for next to nothing. However, in such a case, you need to make sure that enough of a feed supply is available to support your critters.

Watch for opportunities like these, and you may be able to get a little (or a lot) more animal for your money than you would otherwise.

Going local or out of state

With the Internet, you can find a breeder or seller of animals with very specific characteristics. Whereas 20 years ago getting an animal from another state or even country was only for the rich and famous, now the average person can arrange the sale and transport of anything from parakeets to prize racehorses. Going the distance can be great if you find something that isn't easily attainable locally, but getting something a little closer to home is almost always a better idea. Purchasing closer to home means the animal is used to the climate, so you know he or she will do okay in your yard.

If you find something in a different state — or even in a faraway part of your own state — you have to deal with transporting the critters. Larger animals are typically shipped in trailers, but smaller ones like dogs can go by air. You may also have to deal with a quarantine time, in which the animals have to stay in a state shelter for a specified time before they're allowed out into the general population.

Before you put down the big bucks for an out-of-state animal, be sure to check with your state to see whether you have to adhere to any regulations in order for the critter to come into the state. Some states have laws about transporting animals across their borders. In that case, be sure you negotiate with the current owner to decide who's going to take care of any necessary shots and exams, and make sure you factor that into the total cost of the animal.

Evaluating Sellers and Animals

Most animal sales go off without a hitch, but there are some less-than-honest people out there. Unless you dream of playing plaintiff on a daytime TV court show, be savvy about the buying process so you don't end up buying a dud — or fail to get anything at all.

Looking for credible sellers

Watch out for scams, whether from individuals or auctioneers. If you're dealing with a private party (as opposed to a business), getting into trouble may be a little easier because the seller can disappear after the sale and leave you with no support. Or sometimes online transactions are less personal because you may not actually meet — you do your business via the phone, and your purchase is shipped to you.

Be careful when you deal with an online entity, whether you're buying animals or pet supplies! Because a not-so-honest person can hide behind anonymity, you can find plenty of horror stories of transactions gone bad. This is not specific to buying animals, but I can't stress being careful enough. Here are some tips for long-distance transactions:

✔ A good way to get information on distant sellers and animals is to check with other buyers (ask for references) or with an animal association. Contact the association to see whether this buyer is listed and whether you can find any comments about him or her. One more place to check is with the Better Business Bureau.

✔ Some online sites have a rating system where buyers and sellers rate their transactions. A seller with close to 100 percent positive feedback is likely to be good to you as well.

✔ Check whether the seller has a Web site — most scammers don't spend the money for a Web site.

✔ You may want to withhold credit card information until you're sure the deal is a legitimate one.

Judging individual animals

Judging individual animals comes down to doing research beforehand so you have an idea of what you're looking for. You can find local shows for all sorts of animals (dog shows, horse shows, and llama and alpaca shows), gobs of information online, and also numerous magazines targeted to the farmer that may help you recognize what constitutes a quality animal.

An actual meeting allows you to see how the animal looks as well as how the critter reacts to people and to being haltered. Here's how to evaluate animals — or get others to do it for you:

✔ **Getting a little help from your friends:** Friends, neighbors, and even the local vet are good places to ask for help with deciding on an animal. Maybe your neighbor has a particularly good animal who has siblings just waiting for you to take them home. Ask the local vet about a particular animal or seller to see whether he or she has an opinion.

✔ **Doing in-depth visits:** Especially with big-ticket animals (those you'll be spending several thousands of dollars for), you may want to make multiple visits to the farm where the animal currently lives and see how the critter and all his or her buddies are doing. Watch to see how they interact with each other and with their humans.

✔ **Holding a quick meet and greet:** Sometimes, such as in the case of an auction, you have only a short time to evaluate the animal, so make the most of it. Go with your gut. Sometimes, an animal just speaks to you, and you will see something in the animal that makes the decision for you.

✔ **Evaluating animals long distance:** Evaluating an animal long distance is challenging, but you can take some steps to be better protected. Get online and take a look at the available animals. Call and talk to the seller and have him or her send more photos — pictures from the front, back, and side and close-ups of the teeth and legs. Videos are nice, too. However, remember that pictures aren't foolproof — dishonest sellers can send you pictures of a different animal or touch up the photos. If the seller shows the animals and has awards, then chances are you'll be getting something decent.

Ask for papers that show the animal's family history. In some cases, family history can tell you whether the parents or grandparents are known for producing good-quality offspring. Jot down the names and go look them up if possible.

To the Market! Finding an Animal Source

Although all the animals you acquire are purchased from people, you can approach some special ones about buying animals. This section introduces the main players.

Before you start looking for an animal to bring into your family, be sure you're ready for the critter to join you. If you start looking and find the perfect deal on the perfect animal but you don't have a fence or a shelter, you'll be frantic trying to get the place in order and still get in on the deal. If you already have some animals and are bringing a new one into the fold, have a holding pen ready to put the new one in while everybody gets used to the changes.

Consulting your friends and neighbors

What better place to get an animal than from a neighbor? You may know the animal personally, or at least you know about how the owners are treating the animals. Besides, your neighbor is just across the fence if you have a question that comes up. Even if your friends and neighbors don't have animals to sell, they can very likely help connect you with trustworthy sources.

Seeking out reputable breeders

Some people find their purpose in life is to breed animals and sell them to discriminating buyers. Reputable animal breeders know everything there is to know about a certain breed and are very discriminating about which animals they allow to mate so the resulting offspring have the characteristics that should be allowed to proliferate. You can find breeders

- Through lists in your local chapters of animal organizations (such as the Alpaca Owners and Breeders Association)
- By word of mouth
- Through an Internet search

Even those breeders in far corners of the country can be well-known as good, conscientious breeders that anyone, anywhere can buy from with confidence.

Some uneducated breeders think their poor-quality animals are the best and try to sell them as such. Buyers need to be educated before buying. You may want to get a second opinion, so ask the vet or other breeders what they think of a particular breeder. Request papers and references, and check out the breeder before you buy.

Finding animals in the classifieds

Maybe people just want a more hands-on experience with placing their animals so they can see and meet the new owners. Classifieds — in a newspaper, online, or even just a note tacked to the board at the local grocery or farm store — are ways to get the word out that an animal is in need of a new home.

Checking out neighborhood bulletin boards

Some feed and tack stores (and even a grocery store in a rural community) have those good old-fashioned cork boards where you can post a note with a push pin right inside their doors. These are good places to find some great treasures. People who maybe don't want the hassle of placing an ad may instead just put up a sign here. When you're ready to get your critters, you may want to start perusing these boards whenever you go into these stores.

Browsing your local newspaper

Check the local classifieds in the paper. Most newspapers have a section for pets and/or livestock. One benefit of going this route is that the sale is personal. You can meet the animal as well as the current owners before the sale, and you may be able to call on those owners if you have questions after the purchase.

Another benefit is that the sale is local — the animal usually doesn't have to be transported very long distances or across state lines.

Surfing online classifieds

Lots of Web sites (local or national) have want-ad sections where you can get all sorts of things, even animals. The national site agriseek.com is geared toward livestock, tools, farm services, and other agricultural categories, and www.craigslist.com is a national site with local subsites that may offer animals for adoption. The national Web sites have local sections, so you can deal with animals and owners in your own area, or you can search across the country for that special animal you can have transported to you.

You can find loads of local online classifieds as well. These days, even your local newspaper likely has an online counterpart where you can peruse the classifieds online.

Visiting county and state fairs

County and state fairs are good places to find animals. During the show, the animals are available for viewing or meeting. These animals have been well taken care of and are highly trained. The cattle and pigs are sold to slaughter, but you can join in the bidding to get a great pet. You can find ducks, chickens, rabbits, goats, and sheep, too. Farmers come to fairs from all over, so fairs can be a really great place to buy your animals if you don't mind competing with other buyers.

Adopting from a rescue group

Sometimes animals get rescued for one reason or another. Perhaps the past owners got in over their heads and realized they just couldn't manage caring for the animals. Or perhaps the animals became displaced due to a death in the family. The animals may end up in a rescue shelter.

Checking with an animal rescue group can turn out to be a great idea, especially if you're looking for a companion animal. You can find some gems for a low cost. However, some rescued animals may have health problems or have special training requirements, so make sure you know what you're getting into.

Bidding on animals at auctions

Rural areas sometimes have regular animal auctions. This is a quick and easy way for people to get rid of unwanted animals for whatever reason. Maybe they have too many, they don't want to raise a certain type of animal anymore, or they're getting out of the farm business altogether.

Often, animals that go to auction are of poor quality, injured, or not healthy. Some big breeding farms may be getting out of the business or reducing their stock, but I often find they're getting rid of the animals that they can't sell. I've seen breeding farms buy mares, breed them to their stallions, and then have a going-out-of-business or reducing-my-herd auction, when in reality the animals they're selling were never a part of their normal breeding program in the first place. Buyer beware!

Animals put up for auction come in all levels of quality, so find out about certain breeds before you go. Have an idea of what to look for in the animal. Go a few times before you do any bidding so you can get a feel for what happens at an animal auction — what other people bid on, how much the animals go for, and so on. Maybe visit the auction several times to see how it works before participating yourself. Here are some tips for buying at an auction:

- **Get there early and take a look at what's on the table that week.** It always helps to go look with somebody who knows more than you. Perhaps a neighbor who has been through the auctions (maybe on both sides of the sale) can give you pointers on what's a good animal and what's a good price.

- **Talk to people while you're there.** You can sometimes pick up some tidbits just by chatting with fellow auction-goers. A casual remark by, say, one of the assistants at the auction can give you some amazing insight. You can discover the history of an animal, what others think is a fair price, or maybe a secret fact about an animal's health.

- **Know what you want to spend and stick to that limit.** It's easy to get caught up in the bidding. Keep a level head. Yes, maybe this is an awesome animal, but animals come and go in auctions every week.

Of course, there are always exceptions. If you know an animal's history and you really want that one, maybe you're prepared to go a bit higher than what you think is a decent price. Splurging is okay, but don't let this attitude be the norm.

✔ **Dress for the occasion.** Attending an auction means you may be out in the elements for a few hours, so you'll be walking around an area that's not conducive to wearing nice shoes or pants.

You need to have a really tough skin to go to auctions. I went only once, and I thought the animals were treated roughly. Animals have to be moved quickly, and sometimes that means rough corralling techniques. It also made me very sad to see some animals not get sold. (If they're not sold, they simply go back to the seller.) If it were possible, I would've taken in *any* animal that was otherwise not wanted, but that's another story!

Some auctions have some not-so-honest auctioneers. If the auctioneer sees that you're not a regular (a greenhorn who maybe doesn't know the business of auctions), he may invent an opponent to get you to keep bidding and bringing the price up. You may in fact be the only one bidding on a particular animal, but he nods to a non-existent bidder behind you (who you can't see) to make you think you're *not* alone in the bidding. It may be a good idea to have a friend or two stand in back of the crowd to be sure this doesn't happen.

Chapter 11

Caring for and Feeding Your Animals

In This Chapter

▶ Keeping them clean

▶ Feeding and watering them well

▶ Dealing with behavior issues

When you have critters, you're responsible for their care. Throwing a bale of hay into the pen isn't going to be enough. Just like humans, animals need nourishment every day and protection from the elements, at least to some degree. You can avert a lot of health problems by paying attention to basic hygiene and care.

Good nutrition, sufficient exercise, dry bedding, good air circulation, and safety from predators all come into play to give them the best chance of having a happy and healthy life. This chapter discusses some of the animals' basic needs: nutritional concerns, protective shelters, and grooming, as well as some behavioral issues.

Taking Care of Basic Hygiene

"Cleanliness is next to godliness" applies to animals as well as to humans. Keeping your animals clean is the best way to ward off diseases and sickness. In this section, you find out how to make sure your critters stay in good shape.

Keep records of actions that don't take place every day and note when they happen. Especially make sure you record things that happen annually, such as shearing, so you know when the next treatment should be. And as you take care of daily hygiene, use the contact to make sure everything is okay and to bond with your animals so you can recognize individual personalities. Monthly, more-thorough checks are a good idea because animals can hide their illnesses, and changes in their condition can be subtle. Chapter 12 discusses some signs of illness or injury as well as possible treatments.

Because animals need to be restrained or corralled at times, such as when they need grooming or shots, you should find out how to properly restrain them. Doing so makes some of the daily tasks involved in their care a lot more enjoyable for the both of you. See "Training animals for human interaction," at the end of this chapter, for details.

No matter how well you care for your animals, some are going to succumb to illness or injury, and how you dispose of their carcasses is a major hygiene concern. I discuss carcass disposal in Chapter 18.

Cleaning house: Maintaining the living area

Probably the most basic way to keep your animals happy and healthy is to give them a nice place to stay. Here are three main aspects to keeping your animal's shelter in good condition:

✔ **Keep it fresh and dry.** Change the bedding regularly and make sure the pen has ample drainage — no big puddles or large piles of doo-doo are allowed to form. Not only is wet and soiled bedding uncomfortable to lie down in, but it's also a health hazard. It promotes bacterial growth in both the bedding and on the animal, which can lead to infections.

Dispose of manure and soiled bedding in a conscientious way. When mucking out a stall (yep, that's the official term, and you only have to clean out a stall once to know why it's called this), you can add the soiled bedding to the compost pile. If your animals are producing way more manure than you can reuse yourself (for composting or direct application), put a sign out advertising that you have some for free or sale, or donate to organizations such as local parks or even local land-scaping companies.

✔ **Keep the shelter well-ventilated.** In the summer, you don't want hot, stuffy air to build up inside and make the animals uncomfortable. But in the winter, too, humidity from urine, manure, and body moisture can accumulate and lead to respiratory infections, even pneumonia. Being continually subjected to the smell from ammonia in urine is bad for the immune system, but it's especially bad for pigs. Thus, pigpens are typically equipped with fans to regulate the air turnover.

✔ **Keep it warm.** During the winter months, you also need to provide extra straw for insulation.

The beastly barber shop: Bathing and grooming

Most farm animals don't get actual baths, but you may notice they frequently take dirt baths. They roll in the dirt to fend off bugs, placing a dirt barrier on the skin. Rolling around also serves as a great back scratcher. However, you don't need to squirt critters with a hose unless they get really icky and dirty or unless you're getting them ready for a show.

Grooming and shearing are sufficient means of keeping your animals' coats beautiful and keeping the animals themselves comfortable and smelling their best. Just as you have to take care of hair, you need to care for animals' hooves and nails, too. This section discusses personal grooming and basic hygiene for farm animals.

Brushing and cleaning coats

Grooming isn't just so animals look pretty. Doing so can keep them more comfortable. This means you should keep their coats brushed and free of stickers and burrs.

Responsible owners don't allow an animal to live with mud and feces caked on. Part of the daily care for some animals, especially horses, involves brushing the coat. For others, such as cows, proper bedding is enough to keep them clean. But whatever the grooming method, a clean animal is a happy and healthy animal.

Shearing hair

Shearing animals allows a good, clean, unmatted coat to grow in, keeping animals warmer in the winter. Shearing in the beginning of the summer can also help keep the animal comfortable through the hottest part of the year.

Fiber animals, such as sheep, Angora goats, and llamas, require a once- or twice-a-year shearing. You may also want to consider shearing other long-haired animals whose hair isn't usable (or good enough quality) for turning into yarn and other fiber as needed to help the animal feel more comfortable. See Chapter 17 for information on shearing your animals.

Trimming toes and checking hooves

Animals who roam on soft ground — grass or mud — need extra work on their feet. Nails and hooves grow just like human fingernails and need to be periodically trimmed.

Failure to care for hooves can lead to serious problems. A friend of mine acquired a cow whose hooves hadn't properly been cared for and had severely grown out so that the cow was nearly lame. He had to continually and slowly trim a little at a time, hoping to get the hooves back to some semblance of normalcy. But after several months, it became obvious that the cow was too far gone, and my friend decided to send the animal to the butcher.

Animals whose pastures contain a lot of rocks have a better opportunity for their nails and hooves to naturally wear down — the rocks act as natural files — so in those cases, you may not have to trim them at all or at least not have to trim very frequently. However, if the animals live on or frequently stand on very rocky ground or pavement, hooves can chip and crack. Animals need their hooves to walk, so *hoof crack* needs immediate attention from a veterinarian or a farrier.

Hooves and pads can also get stuff in them, and you have to clean them out for the animals. See Chapter 6 for the tools you need.

Horseshoes: Farrying feet

Horse hooves are made of keratin (the same substance that hair and fingernails are made of), which is a relatively soft material. Heavy loads on hard surfaces are a recipe for hoof problems. Improper hoof care can lead to infection and lameness, so find a reputable *farrier* in your area to take care of the shoeing and trimming (see Figure 11-1). If you choose to acquire the tools and do this yourself, make sure you get adequate training.

Figure 11-1:
A farrier at work filing a horse hoof.

A horse needs horseshoes if you're going to have him or her carry heavy loads or plan to have the critter spend a lot of time on hard surfaces. Typically, horses are acquired to be ridden, and a full-grown man can be a very heavy load. Oxen that you use to pull heavy loads may also benefit from shoes.

However, not all horses get shoes. With the proper hoof trimming, horses who don't carry loads or walk around on hard surfaces usually do okay without any shoes. For information about natural hoof trims, ask your farrier about the *barefoot method*.

Letting them romp

Animals don't like to be cooped up, and subjecting them to this treatment can affect health and temperament. Have you ever watched a horse or goat running and jumping so that they seem to be giggling? Running is fun, and being in the fresh air and stretching their legs and chasing their friends around the pasture isn't just enjoyable — it's therapeutic!

Busting loose is a year-round activity. Just because it's cold or snowing outside, you shouldn't assume the animals don't want to go outdoors. Although the animals seek the shelter of their enclosure at times (during sideways snow comes to mind!), most animals really like and even prefer to be in the fresh air, and they generally go outdoors (at least for a little while) even in the harshest weather.

Proper exercise depends on the animal:

✔ Horses who live in stables need the most attention. Riding them regularly is best. Otherwise, you can consider getting those circular motorized walkers that look sort of like a merry-go-round. You hook the horse up to it and turn it on, and it walks the horse.

✔ Llamas tend to run around at times and don't really need anything further, unless you want them in hiking shape. Then you want to regularly take them for a spin around the neighborhood.

✔ Cows, especially those you're raising for meat, don't need a lot of exercise because that means weight loss.

Feeding Your Animals Right

You can't typically find food for your animals in the grocery store (although the pet aisle *can* have some things like Purina Rabbit Chow!). Your best bet is to seek out a farm or country store. Local independents and national chains — such as IFA (Intermountain Farmers Association) Country Stores, Mills Fleet

Farm, Tractor Supply Company, Blain's Farm and Fleet, and others — can undoubtedly help you with a good part of the things you need, especially the feed that comes in bags and the supplements and salt licks. Also, you can find some supplies on the Internet and have them shipped anywhere in the world.

Hay, however, you have to find locally. Look in the classified ads in the local newspaper (under *hay and alfalfa* or *livestock*), find sources through word of mouth from your neighbors, or check message boards at the country stores. After you find a source, try to work with the same people next time. You'll be considered a good customer, and they'll be more apt to work with you during emergencies (like when you run out of hay even though you were meaning to call to order more).

Storing feed properly

Buying small quantities of animal feed isn't very efficient, so buy a bunch of it at once. If you don't properly store the feed you get for your animals, weather can destroy it or the local wildlife can eat it up, so you need good storage facilities for all that food.

Hay, at the very least, should be covered with a tarp or even stored inside a barn to keep the weather and wildlife away. Other feeds, such as minerals or treats, also need proper storage; these containers can be as simple as plastic tubs. See Chapter 5 for more information on places to store feed.

If hay has too high of a moisture content (more than 22 percent), bacteria can cause a chemical reaction that produces a lot of heat. Too much, and the haystack can spontaneously combust, which can ignite your barn and spread fire to nearby buildings or cause brush fires. Don't pack the bales closely together, and leave some air pockets in your haystack. Good circulation greatly reduces the possibility of fire. Keep an eye on the hay for the first six weeks after baling — temperatures above 150 degrees are dangerous.

Finding the right diet

The right diet is important for a healthy life, and each animal has individual dietary needs. Table 11-1 can give you the basics, but within a category, you may have animals who need more or less than the average or may require supplements. Your vet, cooperative extension service, and neighbors can give you more-specific advice.

Table 11-1		Typical Foods for Livestock		
Animal Type	**Main Food**	**Daily Amount**	**Treats**	**Extra Needs**
Camelids	Pasture, browsing materials (twigs, prunings), and grass hay	2% of body weight	Grain, apples (no core), vegetables (lettuce, carrots), fruit scraps	Trace mineral mix formulated for llamas; you may need llama supplements
Cattle	Dry matter or grass hay; alfalfa is too high in protein — use it for sick animals only	2% of body weight	Beet pulp (commercially made beet pellets), maybe mixed with a little molasses; avoid grain — it's too high in calories	Salt licks
Chickens	Feed purchased alfalfa, cabbage, at farm stores	Chickens tend to self-regulate food-intake, so fill a bucket and refill when it's empty	Greens such as fresh pasture, alfalfa, cabbage, lettuce, grain sprouts, and Swiss chard	For laying hens, opt for a feed with a higher protein mix to pro mote egg pro duction
Goats	Grass hay (no alfalfa) and pasture grass	2–4 pounds of hay per day	Some browsing materials (prunings, wild grape vines); avoid grain — it's too high in calories	Salt and copper mixture
Horses	High-nutrition feed, feed, such as grass, clover or alfalfa hay	Varies depending on what they're used for (work pet, show); an adult doing light work generally gets 2% of body weight	Grain, apples (no core), carrots	On average 50-60 grams of of salt per day; lactating females or high-performance horses need a higher-protein diet

(continued)

Table 11-1 *(continued)*

Animal Type	Main Food	Daily Amount	Treats	Extra Needs
Pigs	Grazing materials (grass, roots or other vegetable matter); commercially bought pig feed, supplemented with alfalfa or grass hay	5–6 of body weight until they reach 100 pounds; 5% of body weight from 100–150 pounds; 4% of body weight after 150 pounds	They love treats especially leftovers; don't give them anything they can choke on such corn cobs or whole potatoes	Salt and trace minerals
Rabbits	Unlimited fresh grass or Timothy hay (or hay pellets) and a variety of fresh vegetables	A minimum of 1 cup packed green foods per 2 pounds of body weight once daily	Carrots, limited amounts of lettuce (too much causes diarrhea no corn, cabbage crackers, beans breakfast cereals bread, nuts, pasta, peas, popcorn, or other human treats	Salt licks
Sheep	Grass hay (no (alfalfa) and pasture grass	2–4 pounds of hay per day	Carrots or weeds from the garden; no grain— it's too high in fat	Copper-free salt licks

Improper feeding can introduce health problems besides the weight loss, listlessness, hair loss, and other problems you normally think of in malnutrition cases. Hoofed animals sometimes develop *laminitis,* an inflammation of *lamina propia,* the internal structure of the hoof. This problem can be related to feed; the diet may not contain enough protein, or sometimes too much freshly cut grass or too high of a grain content can do it. Keeping a good eye on the hooves can help you spot problems like this before they get out of hand.

Supplementing with salt licks and mineral sources

Most animals need some sort of mineral supplement, depending on the breed and on the area and quality of feed. Most animals benefit by having a salt lick nearby (country stores have salt blocks and specially designed holders for them), but some animals also need other minerals that aren't in their main feed. For example, goats need a small amount of copper in their diets. Because hay doesn't contain copper, you need to supplement their diet with a mixture you find at the farm store.

Sheep's livers can't handle much copper, so be careful not to leave copper mixture lying around if you have sheep and goats in the same pasture. If a sheep gets too much copper, that copper breaks down the blood cells, clogging up vital organs and leading to kidney or liver failure.

Trace mineral blocks or grains are also available and, depending on the animal and the soil in your region, you may need to get some. Some areas of the country have soil that's selenium-deficient, so the hay that grows there doesn't have sufficient quantities of this mineral, which aids in muscle health; in those cases, get a salt block with selenium. Check with your local country store, neighbor, or cooperative extension service to see what — if any — you may need. Soil tests can also reveal deficiencies.

Indulging your animals with treats

Who doesn't love treats? Animals are no different from humans in that respect. However, just as chocolate probably shouldn't be one of your major food groups, animal treats are something critters should eat in moderation. Give treats occasionally, not as a major part of their diet.

Handing out treats is more of a social or bonding exercise than it is for nutrition. Seeing the animals running toward you when you approach them with a bucket — one they know from past experience has grain in it — is rewarding. You know they're coming for the grain, but it still warms the heart to think that your animals love you so much!

Treats come in many forms. You mostly give animals either grain or fruit and veggie scraps. Read on for details, and see Table 11-1 for some specific suggestions and restrictions.

Grains

Grain comes in so many mixtures! Barley or oats are the most common, but you can also mix in corn, molasses, and all sorts of other things. Your local country store has some basics, and you can ask the workers there or your neighbors for recipes for mixtures.

Grain is higher in carbohydrates, and if given too much, animals can put on too much weight. However, for an animal who needs to beef up (he's been sick or you want to fatten him up before the trip to the butcher), a diet high in grain can pack on the pounds.

Giving a grain treat has another benefit. Almost everybody gets downright *loving* over grain. I feed my llamas an oat and molasses mixture, and you'd think they died and went to llama heaven when they get to munch on it! A llama or alpaca who normally doesn't want you to touch her will be like putty in your hands if her head is stuck in a bucket of grain or if her nose is snarfing up grain from the palm of your hand. You'll most likely be able to handle her easily — maybe even give her a check-up or just hug her neck!

Fruits and vegetables

Some fruits are good treats, and some aren't. Apples are a big favorite with animals such as horses. Some of my llamas come running for apple wedges, and others turn their noses up at them. I've tried other fruits with varying degrees of success, but I've never been able to entice any of my animals to do anything beyond sniffing a banana.

Animals also love certain veggies — lettuce is popular, as are carrots; cut the treats into smaller pieces so everybody gets a sample and so the littler ones can participate in the joy. Cows and other animals love crushed corn. Try various veggies to see what your crew appreciates.

With any fruit or veggie, try not to give critters the cores, which can present a choking hazard.

Providing a Continuous Source of Clean Water

A good, reliable source of clean and fresh water is essential to any animal's well-being. Just putting out a bowl or bucket of water isn't always enough. Many farm animals step in, sit in, roll in, or even throw their water bowls in the air. Thus, getting a specialized livestock-and-poultry water bowl, bucket, or system is a good idea. Automatic, heated waterers are particularly helpful during the winter (see "Preventing freezes" later in this chapter). In this section, I discuss watering options.

Considering common sources

Some properties are fortunate enough to include natural water sources, but many people need to bring water to the animals. You can get pretty creative with how you keep your animals hydrated.

Streams

A natural stream running through your property can be an awesome water source because you don't actually have to do anything to bring the water to your animals (unless, of course, you're in an area where the winter season means frozen water).

If you do have a stream on your property, make sure that not only are the corrals (fences) set up so each separate pen has access to the water but also that barriers are set up so animals can't go between pens. Use fencing or gates that go down to the bottom and prohibit any animal from sneaking under and entering another (forbidden) pen.

Watering from streams can cause some problems. Animals typically don't know anything about *E. coli* and *Giardia*, so they're not careful about leaving the stream to urinate or defecate. A fast-moving stream can reduce the problem. And without testing, you can't always be sure just how safe the water is — do you really know where it's been? Animals' trampling on the banks of the stream can also aid in erosion.

Stock ponds

Some properties have natural ponds that keep the animals constantly supplied with water. If you don't have a natural pond on your property, you can always build one yourself (see Chapter 8 for details).

The biggest problem with a pond — a relatively static water supply where the water sits and doesn't naturally get refreshed — is that that water can get stagnant and mucky. Dirt, leaves, and feces can accumulate, and before too long, not only is the water not good to drink, but it's also a breeding ground for bugs and germs.

If you want to use a pond for stock watering, get it tested. You may not be able to tell how good the water is simply by looking at it — a nice, clear pond with no muddy residue or algae may still be contaminated. No, ranchers of old didn't do any of these newfangled tests, but ranchers of old didn't encounter today's pollution problems.

If the water does get the stamp of approval, here are some options for getting pond water to the critters:

- ✔ Let the animals wander down to drink at will. With this option, limit access points so the critters don't trample the entire bank and speed up dirt accumulation. Standing in cold water is wonderfully refreshing to a hairy beast when the temperatures are hot, but it compromises the quality of the water, so pay careful attention to keeping the water clean.

  I don't recommend letting the animals drink straight out of streams or ponds. With all the chemicals used in today's farming and salt on the roads in the winter time, let alone animal feces, contaminants get into the water supply.

✔ Use an automatic waterer to pump water from the pond into a trough (and maybe even a small swimming hole) and prevent the animals from actually walking into the pond. You can have animals pump their own water with a *nose pump* or pump it for them with an electric or solar motor.

✔ Allow gravity to pull water down into your troughs.

Wells

Water wells are a great source of water for your animals. Use the well to supply the water to an automatic waterer the same way the well supplies your house. You can simply use hoses or install underground lines directly out to the animals.

Some well water is incredibly nice — fresh and pure. However, other water contains unwanted minerals that you may need to filter out. (See Chapter 2 for more information about wells — getting a permit, digging, testing, and so on.)

Manually filled buckets and troughs

The old-fashioned way of supplying water involves manually filling some sort of container, such as a bucket, trough, or large drum (see Figure 11-2). You may want to use this method when you have to separate animals and you don't have an automatic waterer where you need to take them. An old bathtub actually works really well as a waterer. You can also use plastic barrels with the tops cut off, or you can make a trough out of a 50-gallon drum cut in half.

Water troughs need to be dumped every week; the troughs, cleaned; and the water, replaced. In the winter, troughs have to be heated. (See the upcoming "Preventing freezes" section.)

Figure 11-2:
A cow
drinks from
a water
trough.

Using automatic waterers

If your water is set up to automatically refill (a pretty cool concept, by the way), a float in the system detects when the water goes below a certain level, triggering the valves to open and fill the waterer back up. The automatic waterer kind of works the way you set up sprinklers to automatically water the garden.

You can automate underground pipes or use temporary lines, such as above-ground drip irrigation hoses. You can even install temperature regulators so that when the water gets to freezing temperature, the heat turns on.

These systems can be electric, solar powered, or even "energy-free." They may be commercially made or do-it-yourself. The station where you configure the system should be in a location that's easy to get to so you can easily make modifications, change batteries (if using solar power), or just check on it.

Estimating water needs

You may have read that in hot, dry conditions, a human should consume at least 1 gallon of water a day. But a mature cow needs between 10 and 20 gallons a day! Other animals may not need that much, but they still need a lot. A reliable source is absolutely necessary. If you don't have an automatic system on your property, you need to regularly fill the buckets to keep your animals hydrated.

Water access points should be located where they're most convenient to you because you're the one who tends to their maintenance. Of course, each pen needs its own water source. I installed a big trough in each of our pens and right near the fence so I can access them easily.

The size of the buckets or troughs depends on the number and type of animals you have in each pen. Water needs vary by season, conditioning, and life stages. An average horse needs 10 to 12 gallons per day, but nursing mares in the summer can need up to 25 gallons per day. Monitor the water to see what your animals need. I recommend having a bucket or trough that covers at least two days' worth of your animal's watering needs.

Preventing freezes

In a lot of places in the country, winter means cold temperatures, and that means the water in the outside water buckets is going to ice up. But the small farmer has some options for keeping fresh water from freezing:

✔ Use a heated water bucket, usually a plastic bucket with an embedded heating coil and plug (see Figure 11-3).

✔ Get a submersible tank deicer to place in your bucket (Figure 11-3).

If the animals drink directly from the trough (it's not a holding tank you draw water from), consider using the heater design that screws into the plug in the bottom of the troughs rather than one that has a cord that drapes over the top; animals can pull out cords draped over the top, which can allow that water to freeze and could mean burning out the unit. (It's meant to be On my farm, I use a combination of these two. We keep the smaller units with the heaters enclosed with the goats in the pen that's farthest from the house. (Yes, we have to run an electrical cord all the way out there.) The pen closest to the house gets a 250-gallon tub with a couple of the submersible deicers placed inside. Not only do we use the 250-gallon drum to let the llamas and alpacas drink directly from it, but it's also the source of warm water to put into the goats' buckets. Works for us!

If you're going to run an electrical cord outside so you can plug in your heating elements, protect the connection from moisture. We plug the heating element into the plug and then wrap the entire connection with lots of electrical tape. That way, no water can get in and cause a short.

Figure 11-3:
A water bucket with a built-in heating element and a submersible tank deicer.

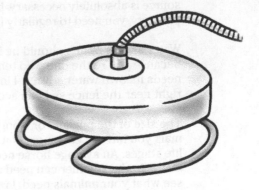

Keeping Your Animals Safe

Animals can't always fend for themselves, so it's up to you to act as their caretakers. This means providing reliable shelter for protection from predators and the weather. Specifics on how to protect animals from predators depends on which animals you have. Rabbits need separate hutches (see Chapter 5 for more info on hutches). Chickens and other small birds need housing to keep them safe. Sheep and goats need a guard animal (see Chapter 9).

Besides these separate shelters and guards, fencing is a must. Not only does fencing keep your animals from straying away, but it also serves to keep larger predators out of the pens, leaving your animals with a little less stress.

Dogs are one of the worst predators of many farm animals, so the fencing should be dogproof (which also makes it coyote- and foxproof). Dogs can get along with your farm animals, but introduce them slowly, and don't leave them alone unchaperoned until you're sure they get along. Dogs can attack and be attacked. Cats usually get along well with other farm animals, except for the baby birds or rabbits.

Never just put a rope around an animal, tether the critter to a fence post, and leave him or her there. A tied animal is vulnerable to whatever predator comes sniffing around, and the restraints prevent the animal from getting exercise. The animal can also get rope burns or even severe injuries from the rope.

Animals should also have a place to go inside when the weather gets really nasty. Some animals don't seem to notice when the weather is bad (llamas sit outside in a nasty storm and act like nothing's wrong), but most animals run for cover, especially when the wind gets bad; therefore, providing a place to get some comfort is a good idea — or even essential. See Chapter 5 for more information on types of shelters you can use.

Reviewing Behavioral Considerations

Each animal has a personality, and sometimes, personalities clash. But besides the differences in your animals' innate personalities, some of the things you do (or don't do) directly affect the way your animals behave. This section discusses some of the behavioral problems that may come about.

Keeping your critters from thinking you're one of them

Animals aren't four-legged humans, so don't treat them as such — even if you have to take special care of a young one. Doing so can cause problems down the road. If you do have to bottle feed an animal, get the little one back with a herd or flock as soon as you can. The adults in the group can show the newbie the ropes.

Bottle feeding typically produces animals who are tamer and more used to the human touch, so they're easier to catch when it's time to shear or trim hooves. However, there's a trade-off. In this section, I discuss some of the hazards of blurring the boundaries between animal-human relationships.

For nutritional reasons, babies should stay with their mothers until they know how to eat on their own. That may be somewhere around two months for some animals and longer for others. (Horses need at least three to four months.) If you're going to give an animal to another farm, one that doesn't have other animals of her kind or age, you may want to leave her with her mother for a couple more months. And also keep in mind that herd animals don't do well if you take them out of the herd altogether. Llamas, for instance, should always go in at least pairs — llamas need social interaction with another of their kind.

Aggression

If you teach a male animal that he's just another person, he may treat people as competition for the females during breeding season, making him very aggressive toward people. If males think you're trying to steal their women, they can attack.

Aggression carries over in other aspects of the animal's life as well. Competition for food can cause the friendly little cow to turn on you if you get in the way of food. After all, he thinks you're one of his kind and that you probably just want to deprive him of some of the food in the trough.

Aggressive behavior manifests differently for different animals. A llama who thinks you're just another llama may have no problems spitting at you if you get too close to food, for example. Llamas typically don't spit at people. (In fact, the only times I've been hit were when I tried to break up a fight or otherwise got in the way of a spitting match between the animals.) But if your llama thinks he's the same as you, you're a prime target.

Having billy goats think you're the same creatures can have another unwanted and downright gross consequence. During the *rut* (mating season), they get stinky with urine. Apparently in the goat world, stinky is attractive, and the stinkiest guy always gets the girl. But if you happen to go out into the pen with grain, for instance, your old pal Mr. Goat may come running, wanting some attention from his old pal, and he'll rub up against you. Besides llama spit, a stinky goat is up there with things you don't want to smell like.

Regardless of the amount of (over)bonding you've had with your goats or sheep, you should never turn your back on the males. If they see you as a threat, they can charge and easily knock you down. Some people have been killed this way — knocked down and either hit their heads or got trampled. Not a good way to go.

Improper socialization

A female who gets a lot of attention as a youngster doesn't typically turn aggressive on you. The main concern is that she won't get properly socialized. If she bonds too closely with humans, she may not work as well in the herd. She wants attention from you as opposed to her peers.

When a young animal leaves her mother too early, the little one may not have had enough time to figure out how to *be* that animal, how to interact with others of the breed. If an animal doesn't get the proper socialization, she may not know how to get along because nobody put her in her place when she got out of hand when she was young. Or the animal may not pick up the things her mom teaches in those early days, such as what to eat or what not to eat. A bottle-fed animal may take longer to figure out appropriate foods or even how to play.

Separation from Mama can affect personality traits. Our goat Annie escaped a personality trait that her twin sister has adopted from their mom. CJ (the sister) has learned to be subservient because her mother is. Annie spent the first few weeks away and thus didn't adopt this character trait. Still, being with Mama is usually the best for a baby.

Dealing with social problems

If too many animals are trying to occupy too small of a space, or if you have trouble suppressing overly aggressive actions of a dominant animal, you'll see some fighting and general meanness. Animals compete for food, water, attention from you, and maybe even the prime resting spots. Here's how to deal with this issue.

Handling overcrowding

The ways around overcrowding are to either thin the herd or make a new pen so the living quarters are a little more spacious.

Establish a few separate feeding places. We have a particularly aggressive female llama who pushes everybody else out of the way during feeding time.

Of course, usually after she eats, there's still plenty for the others, but in the case of grain treats, she'd devour the whole bucket if she could. Multiple feeders, even in multiple areas around the pen, are in order. Problem solved.

Removing weapons

Goats typically use their horns in fighting and to establish the pecking order, but the horns can cause injury to other goats (and humans), so a common practice is to *disbud* them (remove the bud or beginning of the horn) when they're three to seven days old. You can also dehorn an older animal, but that process is more involved, requiring a sedative and a veterinarian.

Castrating aggressive males

One method to reduce aggression in male animals is to have them castrated. Although larger animals have to be taken in to the vet, smaller animals such as goats can be castrated by placing bands around the testicles at an early age. The blood supply is slowly cut off, and the testicles shrivel up and fall off. This method is easier and less expensive than surgery, but you do face certain risks. Never use old, brittle bands that can break, because if the process has started and the band breaks before it's complete, you can lose the animal to gangrene. Make sure you get some training before you try this method.

Separating aggressors

You may encounter an animal who's particularly aggressive toward some or most of the herd. Or you may have an animal who's simply mean and doesn't want to get along. In males, it may be a testosterone thing, and having him gelded — castrated — can fix the problem. Or you can experiment with different roommates until you find a good combination.

We have one male llama who's a good breeder, but he went bonkers when we introduced females into the herd, attacking the other males in his pen. We separated them so he was alone, hoping he'd calm down as soon as he got used to the females across the fence. After a few weeks, we put what used to be his best buddy back in with him. He immediately attacked. We did the process again (separate, wait a few weeks, reunite). It still didn't work. We'd rather not geld him, so we plan to leave him in his own area of the pen for now, hoping that maybe in time, he will settle down. Maybe after a while, if he gets lonely enough, he'll accept a male back in with him. This separation isn't ideal, but we feel it's better than subjecting animals to injury.

Llamas are herd animals and should never be alone, but it may be okay to have one uncooperative male in an area by himself as long as he can see the rest of the herd. Ideally, all animals of the same gender should be able to socialize in the same pen. That's how they're happiest. But as long as the llama can see his herd mates, he won't be too isolated. He still feels part of the herd, even if he's in his own world.

Training animals for human interaction

Unless you're going to be taking the animals to shows — or if your special little gal has designs on becoming a movie star — most farm animals don't need training. You need to be able to control them somewhat to get them to go into holding pens, for instance, but most animals are either small enough (goats and sheep) or otherwise naturally follow the herd so you can get them to go where you want them.

A couple of exceptions are horses and llamas (if you intend to use them as pack animals). Training is essential with these animals because they're large and because you need to get very close to them to ride them or to put a pack on their backs. At the very least, you need to halter-train them so you can lead them around. They aren't born with the instincts on how to act when you touch them or mount them, and if untrained, they may panic and hurt you.

The earlier you start training animals, the better. As they get older and stuck in their ways, it'll be harder to train them to do what you want. Starting in on the training early is also a good way to ensure better handling for mainte-nance tasks such as grooming or medical attention.

Chapter 12

Critter Care: Providing Medical Attention for Your Animals

. . . .

In This Chapter

▶ Covering the animal first aid basics

▶ Knowing when to call the vet

▶ Preparing for the pitter patter of little hooves

. . . .

*J*ust as you have to perform regular maintenance on your car (or your own body for that matter), doing routine maintenance on your animals also makes a lot of sense. It keeps them healthy and stops any potentially bad problems from getting beyond the point of fixing. You can head off a lot of trouble simply by keeping things clean and taking care of your animals' basic food, water, and shelter needs, as I discuss in Chapter 11.

However, you still have to provide medical attention from time to time. Sad but true, animals do get sick, or they get into fights, or they have difficult births. Doing some of this care yourself is easy — and preferable — because calling a vet, especially if he or she has to come out to your property, can be pricey. Or a vet visit may require a wait, and if the problem is something minor, putting on your doctor's cap and administering the treatment yourself is okay. For other procedures, you want to call for backup.

Deciding whether a problem is bad enough to call the vet can be difficult. In this chapter, I note what you can do yourself, point out some of the common ailments that are most likely to crop up, and also discuss a few animal-specific problems. *Remember:* These tips are only guidelines — consult your vet for specific advice. (Check out *Horse Health & Nutrition For Dummies, Dog Health & Nutrition For Dummies* [Wiley], and other animal titles for more medical info.)

Medicine involves many subtleties. For instance, a simple eye infection can look like a scratch; treatment for a scratch when the problem is actually an infection can lead to loss of the eye. You can check out the *Merck Veterinary Manual* (www.merckvetmanual.com) if you want to read more about animal ailments, but unless you've had plenty of training, don't try to self-diagnose.

Putting Together Your Animal First Aid Kit

Minor problems and emergencies are a fact of life, so you should have a few supplies on hand to deal with them. If you plan to keep animals on your farm, having a first aid kit for them is a must. In this section, I give you an idea of what to include. Keep it wherever's most convenient for you, possibly inside the house to ensure it's free from critter pilferage.

Here's where you can get your supplies:

- ✔ The best place to get first aid supplies is likely your local veterinarian — this is my first choice because vets tend to know all the ins and outs of animal care.

- ✔ Try a farm or country store. In the Southwest, a chain called Intermountain Farmers Association (www.ifacountrystore.com) has just about everything you need, and Mills Fleet Farm stores (www.fleetfarm.com) in the Midwest are similar.

- ✔ The Internet has made the world a little smaller, and with just a click of a mouse, you can now easily find products that once weren't so easy to get. Country stores may sell products online, or you can check out a medical supplier such as United Vet Equine (www.unitedvetequine.com).

Including creams, antiseptics, sprays, and meds

Accidents and injuries happen. Sometimes you can easily take care of them yourself, so keep a basic first aid kit around. You should have at least these items:

- ✔ **Antibiotic cream:** You use this for minor wounds to remove bacteria and prevent more from growing.

- ✔ **Bag Balm:** Its intended use is to keep cows' udders soft (thus its name), though it's really good on anything that's dry and cracked, such as noses, calluses, or foot pads. Some cosmetics distributers sell it as a moisturizer for humans.

- ✔ **Hydrogen peroxide:** Because of how it foams up, hydrogen peroxide does a good job of getting dirt out of wounds. It can also be used as a *disinfectant* (it sterilizes inanimate objects) or as an *emetic* (it makes animals vomit) in case they get into rat poison.

- ✔ **Iodine:** An iodine solution (usually 0.5 percent iodine) is a good topical antibacterial solution.

A stronger solution (7 percent iodine, also known as *tincture* of iodine), is not recommended because it can be very irritating to the skin. You don't want to get any in the eyes for sure.

✔ **Scarlet oil:** Used primarily in horses or mules, this stuff is a nondrying topical antiseptic used to promote healing in the treatment of cuts, wounds, abrasions, or burns. You can find this stuff in the local farm store.

On horses, my vet uses scarlet oil for anything above the knees and a light iodine solution for below the knees.

✔ **Gentle antiseptic solution:** Some people prefer to use antiseptic solutions such as chlorhexidine (Nolvasan) on wounds and cut umbilical cords because these solutions aren't as irritating as iodine or hydrogen peroxide.

✔ **Miscellaneous medications:** Deworming meds or vitamins can help. For allergies, maybe include some diphenhydramine hydrochloride (Benadryl), which is used mostly in small animals — such as dogs — who get stung by wasps or bees.

✔ **Fly repellent spray:** Animals attract flies — it's a fact of life on the farm. In many cases, the fly problem can be worse than whatever smells the animals make. Dealing with flies is a big job in the hot summer months.

Fly prevention is particularly necessary around sick or injured animals. Flies get into wounds and lay eggs, which hatch out into a really disgusting creature — the maggot.

Gathering basic equipment

Your first aid kit also needs some basic equipment for monitoring temperatures, cleaning and covering wounds, and otherwise taking care of hygiene. Here are some tools of the trade:

✔ **Basic bandaging material:** You typically use bandages on extremities — after all, you'd go through an awful lot of bandages if you were to wrap them around the animal's chest! Wrapping up a wound on a leg, neck, or tail can really help speed recovery. Besides protecting wounds, you can also wrap bandages around something to keep flies off.

✔ **Clean cloths:** Use clean cloths to create hot packs for treating injuries around larger body parts, such as an animal's torso. See the later section "Treating minor scrapes and cuts" for details.

✔ **Gauze pads:** These come in a variety of sizes (typically 2 inches, 3 inches and 4 inches square) and with different plies (the more layers, the more they soak up). Have a supply of several different sizes and plies on hand for whatever comes up.

✔ **Long cotton swabs:** Useful for cleaning tight spaces such as little puncture wounds, abscesses, and occasionally an ear (you may actually be able to remove ticks), cotton swabs are handy to keep around. But use them with care — if you don't have a good hold on the animal, you can cause more injury.

✔ **Syringes:** Keep a few on hand — they come in different sizes. Use them to flush solutions (such as hydrogen peroxide) into a wound. You can also use these to feed newborns or to give oral liquid medications.

✔ **Bulb syringe:** A bulb syringe is great for sucking fluid out of a newborn's nose and mouth to help the little one breathe on his or her own.

✔ **Thermometer:** Thermometers are, obviously, for taking temperatures. Make sure you label it *rectal* (once a rectal, always a rectal). Choose from the old-fashioned glass kind that usually takes about a minute to register the temperature or the newfangled digital kind that gets a reading in a few seconds.

✔ **Fly trap:** The more flies you can trap, the fewer you have buzzing around your animals and breeding more little pests. And I don't mean using those strips with sticky stuff on both sides. You need something more heavy duty, such as the Ridmax fly trap in Figure 12-1. Or order some fly parasites — tiny, wasp-like insects that eat fly pupae.

For those times when the animal has to go to the vet, you also need some sort of transport. Smaller animals such as goats or sheep can ride in a pickup truck with a cab on it or inside a large dog kennel. For larger animals, you need a trailer.

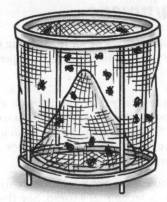

Figure 12-1:
This type of flytrap, which is very useful in barns, traps more flies than flypaper.

An Ounce of Prevention: Scheduling Vaccinations

Every species has different preventive vaccination needs, with emphasis on the word *needs*. If you don't get regular vaccinations for your animals, sickness and even death can be just around the corner. Although you can't possibly vaccinate against all diseases an animal can get, each animal should get at least a few basic vaccines. Find a vet with experience with farm animals and schedule regular visits. In rural areas, you may be able to find a mobile vet, one who comes to you.

Good hygiene, up-to-date vaccinations, and all that preventive stuff pay dividends in money and time down the road. Routine maintenance and attention to minor problems can help keep your animals comfortable and happy.

Knowing the basic vaccines

Just as humans receive childhood vaccinations, animals should get vaccinations against some diseases as a matter of course. Which diseases you have to watch out for depends on what's prevalent in your area, the history of your herd, and whether a disease is likely to be contracted. But as a general rule, consider the following vaccinations:

- ✔ **Tetanus:** All animals can get tetanus, but horses are particularly prone to it. The tetanus bacteria is in soil and is usually introduced into animals via a puncture wound — from nails, barbed wire, or any kind of deep cut. Tetanus causes *lockjaw* (their jaws actually lock up so they can't eat), convulsions, and sensitivity to light.

- ✔ **Clostridial diseases:** Animals can catch all sorts of diseases from *Clostridium* bacteria, including foot root, botulism, and intestinal diseases. (Tetanus is actually a clostridial disease.)

- ✔ **Rabies:** All animals are susceptible to rabies. This nasty virus is transmitted by bite wounds, and it often causes death. It has two types:

  - • *Dumb rabies* causes the animal to become dull and lethargic and not want to eat anything.

  - • *Furious rabies* causes aggressive behavioral changes and foaming at the mouth.

- ✔ **Brucellosis:** This disease causes stillbirths and miscarriages in cattle and is also transmittable to people.

The human form of this disease, called *undulant fever,* is a flu-like disease with symptoms that never go away. It's incredibly hard to cure, and the infected human usually has a chronic flu, exhibiting symptoms every couple of weeks.

✔ **Mastitis:** Animals who produce milk (cattle or goats) can get *mastitis,* an infection of the mammary gland (the udder) that causes clotted milk or a hard or tender udder.

✔ **Nasal infections and pneumonia:** Pigs should be vaccinated against these.

Young animals are often more susceptible to infection and are thus more likely to benefit from certain vaccinations that may be unnecessary when the animals get older. For instance, young animals may need to be vaccinated against respiratory diseases.

Consulting your vet for immunizations

Not all vaccines are alike. Some may not induce a good enough response, or they may generate unforeseen reactions. So in terms of the health of your animals, your local vet is usually the best qualified to evaluate vaccines and choose to use the one that's likely to have the best results. How often your animals need vaccinations will vary — your vet can help you set up a schedule. Here are some other reasons the doctor is likely the best one to be administering these (and any) meds:

✔ In very rare cases, animals have bad reactions to vaccines. If you give the vaccine yourself, you may not pick up on the reaction, or you may not get your animal into the vet in time. One such reaction is severe anaphylactic shock, which causes difficulty breathing and a sudden drop in blood pressure — the animal can just collapse. Death can occur within a very short time, so getting help fast is critical.

✔ The drug needs proper handling. It should be refrigerated, diluted correctly, not exposed to heat or freezing temperatures, and administered with a sterile syringe.

✔ Some vaccines go under the skin *(subcutaneous),* and some go in the muscle *(intramuscular)* — you should take care not to get those vaccines in the vein. If given the wrong way, the vaccine may not be effective at all, or it may cause serious consequences or even death.

✔ Manufacturers don't (and can't) guarantee that all animals given their vaccine will be able to produce the required antibodies to fight off the disease the vaccine was formulated against. So although you think you've given the vaccination, it may not take. Vets have experience and know which vaccines are best.

> ✔ Syringes and needles are considered hazardous waste and must be disposed of in specific ways. You can't just toss the used needle into your wastebasket.
>
> ✔ A licensed veterinarian has to administer the rabies vaccine. If you give it yourself, the state considers your animal to be unvaccinated.

You can get an expert to do the vaccinations and still do it affordably. For instance, you can time the vaccinations so you're doing all the animals at the same time. You pay a base fee for the vet to come out to the farm and then a charge per animal. There may even be a break when you get to a certain number of animals.

Deciding to do the vaccinations yourself

Although most veterinarians don't recommend that you do vaccinations yourself (because of all the reasons I list in the preceding section), many farmers do vaccinate their own animals — and this decision mostly has to do with money. If you have 40 sheep, paying a vet to vaccinate them all can cost a pretty penny. And some animals that aren't very expensive themselves may not warrant spending so much on vaccinations — the house call for the vaccination may cost more than the animal is worth. This isn't a good mindset, but it's one that some farmers get into.

Most veterinarians are willing to show farm animal owners how to administer shots, so consult with your vet before undertaking the task yourself.

Responding When Your Animals Are Sick or Hurt

Despite all your attention to their health, animals sometimes get sick or hurt. Your animal may have an encounter with wildlife and end up losing. Or an errant neighborhood dog may get into a ruckus with your goats. Or a minor medical problem can turn into a bigger one. In this section, I discuss signs of sickness, treatments you can use, and when to call in the vet.

Watching for signs of sickness

You can't know what's *abnormal* until you know what *normal* is. When you go out to feed your animals, give them a once-over to check for health problems. Keep a close eye on them so you catch the minor scrapes and take care of them before they become major infections. Daily observation tells you the animals' habits, what they like, and how they react to things.

Determining whether your animal is just having a bad day or really is sick and in need of attention isn't always easy, but a few red flags always mean it's time to take action:

- ✔ **Limited activity:** If you notice an animal is down, that doesn't necessarily mean there's a problem. He or she may just be resting or basking in the sun. But if an animal is down for a while, try to get the critter up and see what he or she does.

- ✔ **Loss of appetite:** Sometimes an animal just doesn't want to eat. Maybe the day's too hot, or perhaps the animal is wiped out and needs rest first. Or maybe the food is actually bad, and the critter just refuses to eat it. But if your critter is continually just sniffing but not eating, something's wrong. *Tip:* A good way to test an appetite is to give oats or other favorite treats. If an animal who usually becomes putty in your hands when oats are offered suddenly isn't interested, the animal's likely ill.

- ✔ **Weight loss:** Animals may get thin because they simply aren't getting sufficient food, but perhaps it's the result of an infection that's draining the animal of nutrients and energy. Weight loss may also be due to parasites, poor nutrition, or a case of bad teeth that need dental care.

- ✔ **Reduction in milk production:** For animals who give milk, milk production is an even more sensitive gauge than appetite. Lactating females should produce plenty of milk. If their milk production is off, something's up.

- ✔ **Other obvious signs of disease:** A lot of other symptoms can indicate that your animal isn't feeling up to snuff. Excessive coughing, a runny nose, diarrhea, or anything else that's just not normal for an animal probably means he or she needs attention.

If you're concerned, call your vet and ask whether he or she thinks you should bring the animal in (or whether a house call is in order).

Wearing the doctor's cap: What you can do

If you see that your animals aren't in the best of health, you can take care of a few tasks yourself. This section covers ways you can treat and monitor a sick or injured animal.

Taking temperatures

Temperatures outside the normal range can indicate many problems. If you see a difference in temperature, it's probably time to call a vet. Here are some possible causes of high or low temperatures:

Low Temperature	*High Temperature*
Shock (from blood loss, trauma, heart problems, allergic reactions)	Infections (bacterial and viral)
Hypothermia (prolonged exposure to subnormal temps)	Heat stroke (hyperthermia) Seizures Drug reactions

Be aware that taking an animal's temperature isn't going to be met with cooperation, so secure the animal first. The thermometer does not go into the critter's mouth! Otherwise, if the animal balks (maybe violently), you can get hurt or the thermometer may break. Not good.

Animals can suck the instrument up into the rectum, so tie a piece of rubber tubing or twine around one end, secure an alligator clip to the other end of the tubing, and clip that to the animal's hair. That way, if it gets sucked in, you can pull it out, and if the animal tries to push it out, the alligator clamp will prevent it from going flying across the room.

Before inserting the thermometer, lubricate it with something such as Vaseline or K-Y jelly. For a glass thermometer, put it about halfway in the animal's rectum and leave it in about one minute (or follow instructions for your particular thermometer). Be sure to clean it after use — get rid of any residue with soap and water first, and then dip it in alcohol for a few minutes to sterilize.

A normal temperature is different for different animals. Use Table 12-1 as a guide when you're checking temperatures, normal or otherwise.

Table 12-1	Normal Temperatures for Common Farm Animals
Animal	*Normal Temperature (° F)*
Camelids	Around 101
Cattle	101 to 102
Dogs or cats	100.5 to 102.5
Horses	100 to 101
Pigs	102 to 103
Sheep	100.9 to 103.8

Treating minor scrapes and cuts

Scrapes and cuts happen. An animal may get caught up in the fencing, or somebody may get a little too aggressive at the grain bucket. A minor scratch isn't going to be a matter of life or death, but taking care of it can mean making the animal more comfortable and can also prevent the minor problem from turning into a major one.

Determining what's minor is kind of a judgment call. Check the size of the injury (the surface area) as well as how deep it has gone. Then follow these guidelines:

✔ If it goes *full thickness* — in other words, it goes all the way through the skin and into the tissue below it so that you can move the two sides of the cut in opposing directions — then it's more serious. You should probably call your vet.

✔ Even if a cut isn't full thickness, it's more of a concern if it's at or near a joint, the eyes, or a vital organ. You don't want to mess with that yourself.

✔ A surface cut or one that's only 1 inch or so long is probably going to be okay and will respond favorably to minor treatments.

To treat minor cuts, clean them with hydrogen peroxide or an antiseptic solution and then bathe them in some sort of antibiotic ointment or Bag Balm to protect them (ask your vet for product recommendations). Then cover the wound with a sterile bandage. Keep bandages dry and clean, and change them every day or at least every other day.

Don't wrap bandages too tight! No bandage at all is better than one that cuts off circulation.

On those spots where you can't wrap a bandage around the body part (such as a torso), wash out the wound with an antiseptic solution or a half-strength hydrogen peroxide solution (one part hydrogen peroxide and one part water). To reduce swelling and decrease blood flow, put a cold pack on the affected area for 10 to 15 minutes (or however long the animal sits still for it).

If the wound becomes infected, soak a clean cloth in hot water and place it on the wound for a few minutes (as long as the animal lets you). This hot pack stimulates blood flow and promotes local immunity. Finally, put some antibiotic ointment or Bag Balm on the wound to protect it from other bacteria. With horses, you may want to consider an antiseptic spray such as Wound-Kote, which medicates and dries.

As long as the cut continues to heal, you're okay. If it gets worse, a visit to or from the vet is in order.

Deworming

In veterinary terms, *worms* aren't the squishy things you use when you go fishing. These are parasites that get into an animal's gastrointestinal (GI) tract from a variety of sources. Mostly, animals ingest them from eating grass or other things off the ground, but animals can also be born with them.

Worms in the GI tract rob an animal of necessary nutrients intended for the animal. The nutrients instead go to the worms. Getting rid of them, or *deworming,* is a routine act that decreases the worm load that's stealing nutrition from the animal. Some deworming medications are injected, and some are given orally. They usually have to be administered more than once within a short period of time so you can be sure everything is cleaned out.

You can obtain deworming meds from the farm supply store and administer them yourself, but check with the vet for specifics.

Giving vitamin supplements

Whether your animal needs vitamin supplements depends on the conditions in your environment and also the health of your animals. Check with your veterinarian, the local cooperative extension service, or the country store for information about what your animals' diets may be lacking.

Here are some common vitamins shots:

✔ Animals sometimes have fat-soluble vitamin A, D, E, and K shots. Mostly, animals get these vitamins from dietary sources, but in some cases, an animal may need an extra boost because the food sources don't contain enough. Vitamin supplements can alleviate many symptoms of deficiency, such as lethargy, a poor coat, or an inability to gain weight.

✔ Vitamin B is an appetite stimulant. Occasionally, you may get a newborn who doesn't seem to want to eat, and this shot can help. Or an animal can get this shot at any time in the animal's life if he or she just isn't eating and needs a little push.

✔ Some sheep or goats get vitamin E or selenium shots subcutaneously (under the skin) at birth to ward off *white muscle disease.* This degenerative muscle disease causes anything from mild stiffness to obvious pain upon walking or even an inability to stand if it attacks the skeletal muscles; it can also cause difficulty breathing, a frothy nasal discharge, and fever if it affects the heart muscles. Check with your veterinarian to see whether the hay in your area is likely to be deficient. After the initial shot, the animal may need a booster.

Before administering vitamin shots yourself, have your veterinarian show you how. Vitamin or mineral blocks, which are similar to a salt lick, are also a good way to get these nutrients into your animals. You can get the meds at the local farm supply store.

Determining when you need a vet

Some situations are just too much for the average farmer to deal with, and you need an expert to come to the rescue. Usually, you can find a mobile vet in your area, but sometimes you may have to load the critter into a trailer and take him or her to the vet.

You can usually find a 24-hour emergency clinic for smaller animals, and rural communities often have a 24-hour number to call for large animals. In some places, the local large-animal vets form a cooperative where each does a stint of on-call duty and then has time off. You can get such a phone number from your vet or from the local farm store.

If you need a vet, call right away. Waiting can mean the animal won't pull through. (For small animals such as dogs and cats, vets can take an hour or two to get to you; in some areas, even longer.) Here's when you absolutely need a vet:

- **Obvious external injuries:** If the critter is bleeding profusely, or if you see a huge gash or lump, or if the animal has an obvious broken bone, go to the vet.

- **Extremely depressed, lethargic, or thin animals:** Hopefully your animals never get to this point because you're watching them daily, but if for some reason you miss the signs and look out one day to notice that something has gone horribly wrong, please call the vet right away.

- **Critters totally off their food:** You should be concerned if, after 24 hours, an animal hasn't touched a bite to eat and isn't even interested in treats. Horses shouldn't even go this long. If they're not eating, watch them continually. If they go down, it could be a sign of colic, so you need to call a vet.

- **Down animals:** An animal who can't get up has usually crossed into the serious category. If you can't get your animal to stand (that is, he or she isn't just basking in the sun), something is drastically wrong — get immediate attention.

If a horse goes down and starts to roll, get him up immediately and walk him. You now face two potential problems — *colic,* which means the horse can't poop or vomit due to abdominal problems, and the rolling, which can cause the intestine to twist. Call the vet and continue to walk the horse till the vet arrives. If your vet has given you some fluxinine meglumine (Banamine, a prescription drug), he or she may instruct you to give the horse a shot to relieve the pain. But beware — a Banamine shot can mask signs the vet needs to diagnose the kind of colic. The problem may be an impacted colon, a ruptured stomach, worms or other parasites, or many other things.

Your vet is only a phone call away, provided you have access to a 24-hour emergency clinic. You can always call and ask whether the symptoms you're seeing are serious or whether you can do some treatment yourself.

As a courtesy to your vet's private life, try to take care of potential emergencies as soon as they arise instead of waiting until you can't put them off anymore. Yes, emergencies do happen — sometimes in the middle of the night — and your vet will do what he or she can to help save your animal. But maintaining a good rapport with your vet is always a good idea, and part of that means not having to call in the middle of the night because you didn't handle a situation during regular hours.

Bracing Yourself for Birthing Season

Yes, babies have been born throughout the ages and usually without any help from humans. But things can and do go wrong, and if you're there to assist in the birth (or the immediate aftermath) and deal with any problems, then animals that wouldn't otherwise make it on their own can have a chance at life. You want to be sure you have all bases covered at birthing time — especially if your animals are prize racehorses or alpacas you hope to sell for big bucks!

Predicting due dates

If you know the time of conception, you can consult Table 12-2 and have a pretty good idea when the baby will arrive. Sometimes knowing precisely when the deed happened is easy because you've put two critters into a separate pen and let them get to know each other better.

Table 12-2	Gestation Periods for Common Farm Animals
Animal	*Gestation Period*
Camelids	11.5 months
Chickens	21 days
Cattle	9 months
Dogs and cats	58 to 63 days
Goats	150 days

(continued)

Table 12-2 *(continued)*

Animal	Gestation Period
Horses	340 days
Pigs	114 days
Rabbits	33 days
Sheep	148 days

When a couple of critters share love behind your back, you can still do tests to determine whether a pregnancy has occurred. Animal pregnancy tests skip straight to the ultrasound, which works the same as with humans in that it sends out a sound wave that bounces through the body and shows up as a picture. You don't have to anesthetize animals — they don't seem to mind, because the procedure is noninvasive. You do have to shave the belly and rub some jelly on it to aid in the transmission. (However, horses have ultrasound tests done through the anus — a procedure most mares are not too happy about.)

Blood tests aren't as accurate in animals as they are in humans, so farmers don't use them to determine pregnancy. One exception: After horses are 120 or more days along, blood tests are pretty accurate. Very accurate urine tests are available now, too. Another difference with pregnancy tests in animals versus pregnancy tests in humans is that animal tests are less accurate in predicting due dates.

Preparing for a birth

In a perfect world, the animals wouldn't need your help at all. And a lot of births do go off without a hitch. The first llama born on my farm happened without my even being there. Our momma llama came to us "probably pregnant," but the previous owners had no idea when the deed had happened. I came home from a day out and found a baby llama on the ground. Mother and child were doing just fine, and I could take absolutely no credit for it.

With a few precautions to keep the mother comfortable, chances are you'll have a successful birth. Even in those perfect scenarios, you can still take some measures to help the mammal mother prepare:

✔ **If you do have an idea of when, put the mother into an isolated area.** Opt for a place under some sort of shelter where she can stay warm, dry, stress free, and away from outside interference. That means no screaming children or barking dogs. Stress causes more adrenaline production, which can actually stop females from pushing. Wind and other weather nuisances should also be lessened or eliminated.

✔ **Keep a good watch on the mother during this close-to-the-big-moment time, and adjust feed as needed.** Watch mommas closely, a few times every day. Installing a video camera can allow you to do your watching while still getting your chores done. You may want to get a halter alarm, which goes off when the animal lies down; this, along with the camera system, can allow you to be in the house (even asleep) right up until the blessed time.

As the time draws nigh, the mother gets restless and obviously uncomfortable — stands up, lies down, and then does it all again. When she starts doing this, she may want or need to go off her feed for about 24 hours. It's getting closer! Watch for contractions, and watch for her to start bearing down, pushing. If all goes well, a little bubble will come out and then a nose or foot.

The best thing to happen is that you don't have to step in at all. Step in only if you do have to. (See the next section for birthing problems.)

✔ **Have plenty of dry straw around.** If the birth happens in the colder months, you want to bring mother and babe inside for a while. My newborn llama was born on March 1, and because our barn wasn't yet built, she spent the first month of her life with her mother in our garage with a space heater.

Following up on a normal birth

Most of the time, you don't have to do anything after the mother gives birth except fawn over the new baby. But even after a perfect birth, you can take some steps to ensure the baby's and mother's safety and comfort.

✔ Dip the umbilical cord in a bit of Nolvasan or other nonirritating antiseptic solution. Maybe even clamp off the cord and permanently tie it off with some string.

✔ Make sure the baby gets dried off. The mom usually does this, but sometimes you need to help. Use some clean straw to rub the baby and get her good and free of the birth goo.

✔ Make sure they're both warm and dry.

✔ Watch to be sure the baby takes her first milk within the first half hour or so. This first milk is known as *colostrum,* a nutrient- and antibody-rich substance vital to the newborn's health.

✔ If you have several animals having babies at the same time, you may want to tag them so you can keep track of them.

✔ You may want to weigh babies periodically to be sure they're putting on weight.

Handling difficulties before and after birth

Despite your precautions, sometimes things just don't go by the book. A baby can be stillborn or breech. The birth can take a lot longer than you expect. Or umbilical cords can get wrapped around the little one. In this section, I discuss some problems with delivery and problems that can arise later on.

Birthing problems: Dystocia

Dystocia is a fancy name meaning "difficult birth," and that can mean anything from a birth that's taking too long to one where the mother is in excessive discomfort to one where the baby is breech (backwards) or even stillborn.

Watch for signs something is amiss. For instance, perhaps the mother's water has broken and she's in labor but nothing is coming out. If a mother goes through this for more than a couple of hours, call the vet because something is really wrong. In horses, you don't even want to wait that long. Horses have been known to rupture the uterus pushing a foal who won't come out. Call the vet if the mare is pushing for ten minutes and nothing is coming out or if the baby is *breech* (you're seeing the baby's butt end instead of the head).

You may also want to elicit the help of a neighbor who may have gone through difficult births in the past and may know what to do.

If the baby isn't breathing, clear the airway so the little one can begin breathing easily on his or her own. A bulb syringe — or a turkey baster — is great for sucking fluid out of babies' mouths and noses so they're able to breathe. Tear any remaining sac off from around the nose. If that doesn't do it, rub the baby vigorously. Ideally, the mom instinctively knows what to do, but occasionally (especially on a first birth if she's particularly nervous), you may have to help. You may even have to do mouth-to-mouth resuscitation.

Mothers' rejecting their babies, or vice versa

For some reasons, not always immediately obvious, a mother rejects her baby. Sometimes you can figure it out, but sometimes you just don't know why. One speculation is that Charles Darwin's theory of survival of the fittest applies here. Perhaps the mother knows that a particular baby isn't strong enough to make it, and she wants to give her energy and attention to the others that she has more hope for.

Perhaps the mother has *mastitis* (an infection of the mammary glands) and nursing is very painful for her, so she turns away from her babies. Or the mother may be sick herself — maybe wiped out from a difficult birth or something more serious that you just don't know about. Or maybe she's just a nervous, first-time mother who doesn't want to or doesn't know how to take on a newborn.

Sometimes the baby is the one who doesn't take to the mother. The little one may be too weak or too cold to get up and nurse. Or the baby may have a cleft palate or other abnormality that prevents him or her from nursing. Sometimes, the baby simply doesn't learn the suckle reflex and thus isn't taking to the usual form of nursing. (In horses, this baby is called a *dummy foal.*)

Whatever the reason, rejection does happen, and you get left with a baby who isn't going to make it on his or her own and isn't going to get any help from the mother. The next section discusses raising the little one yourself.

Raising orphans or rejected babies

Mothers can die in childbirth or shortly afterwards from non-birth-related accidents, and then you're left with a newborn who needs constant attention if he or she is to have a chance in life. Or you may have an infant who for some reason isn't able to nurse from the mother.

Babies have to be kept warm and fed regularly. You have to have the right facilities and the time to devote to a very needy individual almost 24 hours a day, so deciding to take on one of these orphans is a big commitment. But when letting the baby die is the alternative, many people feel adopting the baby is the only reasonable choice. Orphans are notoriously hard to raise, but doing so can be very rewarding. When you've done the mothering job since day one, you and the animal develop a special bond.

But the bad news is that animals without a mother of the same species can be robbed of natural immunities and thus grow up to be less hardy. The first 12 to 24 hours of an animal's life are critical for several reasons. Besides the obvious reasons of feeding and protection, this is when the mother produces *colostrum* — a thickened, antibody-rich milk that's important for giving the baby a natural immunity. The mother actually can produce colostrum for a couple of days after birth, but the baby can absorb it only during these first hours.

If the babies *do* get that colostrum from their mothers during the critical first hours of their lives, they're blessed with a passive immunity, meaning they're more apt to fend off minor and common illnesses. This immunity from their mother's milk tides them over until their bodies can produce their own immunities. After the important absorption of colostrum, a baby can be successfully raised by a member of another species — such as you! Without it, their immune systems are compromised and they're susceptible to all manner of infections.

Birth defects

Animals can be born with any number of birth defects. Some mean the animal can't possibly survive, some mean the animal will be handicapped but will still be able to live a decent life, and some mean the baby just needs a little help. Common birth defects include the following:

✓ **Cleft palate:** With this condition, the animal can't successfully nurse and instead inhales milk when he or she is trying to suckle, causing a condition called *inhalation pneumonia*. Only surgery can fix this condition.

✓ **Umbilical hernia:** With this condition, the body wall doesn't close correctly over the umbilical cord, leaving a hole in the abdominal wall. Some smaller openings can close up on their own, but if they're bigger than the size of a quarter, they need veterinary attention.

Some less-common abnormalities include two heads, multiple limbs, or even a missing anus. Many other defects are possible, and if you experience one, it's crucial that you consult a vet to determine whether the animal has a chance.

Deciding to give one up

Sometimes on a farm, you have to make tough decisions that you wouldn't have to make if the animal were your only cherished pet. If you want a successful farm, you have to let some of the animals go. Here are some reasons you may not be able to keep your little ones:

✓ Sometimes, you simply have too many animals. If you allow animals to breed, your flock is going to grow, and there may come a time when you get to the limit of what your acreage can handle. Perhaps you bred more animals than you expected to and you simply can't feed or accommodate all the resulting babies.

✓ Sometimes animals grow up to be too aggressive. A particularly aggressive animal probably isn't placeable somewhere else unless someone's willing to devote a lot of time and love to try to change the bad behavior.

✓ Sometimes, a baby just isn't responding to all the help and love you're lavishing upon him or her.

Hopefully, you can find a good home for the ones who are placeable. But in many instances, an animal (who didn't do anything wrong — perhaps he's simply part of your overpopulation problem or he lost his mother) may have to face a less favorable fate. Making the decision to send an animal to the butcher or to adopt him out to another farm is especially hard when the animal is one you nurtured from birth.

Part IV

Choosing and Growing Plants and Crops

The 5th Wave   By Rich Tennant

"That should do it."

In this part . . .

Some things grow on their own. Just look at all those weeds and the grass that has to be mowed, despite your doing nothing at all to help them (in fact, you're probably putting in a bit of effort to make sure they don't grow). But other plants need tender loving care to coax them to grow into something truly tasty to eat.

Before you get started, you have to decide what you want and how much of it to plant. Make sure you get just the right amount — being overgrown can turn you into a burden on friends and neighbors with your overabundance of vegetable gifts. A zucchini is a terrible thing to waste! This part tells you what you need to know to get started and balance everything out.

Chapter 13

Deciding What and How Much of It to Grow

In This Chapter

▶ Deciding how much should you get into

▶ Understanding plant varieties

▶ Putting plants in the right place at the right time

▶ Growing vegetables, fruit, nuts, and herbs

F ew things are better than fresh fruits and vegetables, and growing your own can be very rewarding. You may want to grow just enough for yourself, or you may decide to grow a little extra so you can pocket a little cash. And if you have animals, you may want to raise crops to feed them.

As you plan your garden, you need to consider how much of your day you want to devote to caring for crops, how long you want the growing season to last, which types of crops you want to grow, what kind of soil you have, and how to supplement the soil so it makes for a perfect environment. This chapter helps you answer some of those questions and describes some of the more common fruits and vegetables and how best to grow them.

Deciding How Much to Grow

Just because you have 20 acres of land doesn't necessarily mean you're interested in taking on the task of planting, tending, and harvesting crops on the whole plot. That's a lot of work for a hobby farmer! But maybe you do want to do it. Maybe you have dreams of raising enough to sell to the local grocery stores and farmer's markets and of lining your pockets with some extra cash. Or maybe you actually have dreams of becoming a full-fledged, full-time farmer, and you need to get your feet wet on a smaller scale before taking the giant plunge.

Before you plant your first seed or seedling, you need to determine how much you want and need to grow to suit your purposes, and you have to think about the time you have available for managing your crop or garden. You may be surprised at how much you can actually grow in a relatively small space. Sometimes, the smaller the space, the more attention you can give, and thus the more yield you get.

Supplementing your grocery needs and sharing with friends

Have you ever added up how much you spend on fruits and veggies per year? Wouldn't it be cool if you could reduce or even eliminate that expenditure? Well, with a little bit of acreage, a little knowledge, and a little work, that's very possible.

As far as the amount of space you need, you may hear all kinds of suggestions. One source I found says with 500 square feet, you can grow 90 percent of a person's annual vegetable requirements; however, your results will undoubtedly vary.

If your intention is not only to feed yourself but also to treat your relatives, friends, neighbors, and coworkers, you need a few more plants. Or you may just be content to share any overabundance that grows from what you've planned for yourself. Many plants produce all their fruit within the span of a few weeks, so if your freezer is getting just a little too full of grated zucchini, share the wealth!

To determine how much you'll need, ask yourself how much your family consumes: How big is your family? Do the little ones even like vegetables? A couple of tomato plants may be sufficient for a family of two who mostly uses tomatoes as a garnish or maybe as an accent in sandwiches, but for a family who loves and eats salads at every dinner, you want to go with several plants. Chat with neighbors, the county extension agent, and the people at the seed stores for advice. You'll most likely go through a few seasons of trial and error before getting the garden size right.

Selling at the market

If you have enough land and time, planting a lot more, with the intention of selling to local markets — or setting up a table at the various farmer's markets that crop up in the summertime — can very well give you some well-deserved pocket money, as well as a little fun socializing with the people who come to visit. To get an idea of just how much you should plant, ask neighbors for suggestions, inquire at your local cooperative extension office, or even chat with farmers at the farmer's markets (which can give you planting ideas for next year).

A lot of rules and regulations come with selling food to the general public, and each state defines its requirements. If you operate as a produce stand — you handle only raw, unprocessed fruits and vegetables as opposed to selling prepared food — you may get away with anything from very limited rules to having a food handler's certificate or specific equipment (such as refrigeration) at your booth.

Going all the way to a subsistence-level operation

Being completely self-sufficient — growing enough to meet your basic needs — requires a bit more work. You have to be diligent about caring for the plants, dealing with pests, and so on, and you'll be spending more of your time out in the field than inside enjoying a warm fire or a cold beer.

Although being able to fend for yourself may seem like a romantic concept, you may want to read about some of the hardships American pioneers and homesteaders faced. And no, I don't mean using buffalo chips for fuel, repairing broken wagon wheels, building houses out of sod, and using oxen-drawn plows, because in this day and age, most of those problems are a thing of the past. But you may still face droughts or insect infestations that may mean little or no food one year. Of course, today, you'll always have a backup in place (a grocery store nearby) to run to if your crops fail.

For a subsistence-level operation, you freeze, dry, or can fruits and veggies to preserve them for the off-season. The freezer is definitely an advantage you have over the homesteaders. Instead of having to dry veggies or smoke meats, you can simply toss things in the freezer and have at them as needed, well outside the growing season. (See Chapter 15 for more about enjoying the fruits of your labor.)

You'll probably have to do a few seasons of trial and error until you find out just how much you need to plant. That means estimating what your family consumes in a year and planting what you think will cover that, adding a little extra as a buffer in case some of the plants fail. You may even consider growing a bit more than you'll consume so you can harvest the seeds, dry them, and use them the next season. Then see where you stand at the end of the winter when it's time to start the growing all over again. If you find that you misestimated, adjust accordingly.

Even with all this calculation, you may not be able to be completely self-sufficient. You may get the plants right, but if you want to eat meat, you may need to find a good butcher (see Chapter 16). Or perhaps you want to do some bartering for something your neighbor grows, exchanging some corn for some okra. Or how about trading some tomatoes for some fresh goat's milk?

Garden Variety: Choosing Plant Types

If you think the grocery store offers you a lot of food options, just wait until you start browsing the seed catalogs and nurseries! *Varieties* are subspecies of plants that display certain characteristics — a particular appearance, flavor, or resilience — and there are hundreds of them, with more varieties showing up every year. All this means that with a little digging (research, that is), you can find a plant that translates into a perfect eggplant parmesan or fresh salsa. In this section, I explain some of what goes into choosing the right variety.

Selecting plant characteristics

What you decide to grow, of course, has to do with what *can* grow in your area. Certain plants simply don't grow in certain areas, and others grow well in one area and not-so-well in another. Your growing choices also depend on personal preference. This section discusses some plant characteristics you can choose from.

Getting in the zone: Climates and hardiness

Some varieties of plants are bred to survive cold weather, and others require a certain number of hours of daylight. You want to select a plant that's well-suited to your area.

Plants are also bred to have different growing seasons. The length of the growing season is important because you want to make sure your chosen plant variety has time to set its fruit before the first frost of the season kills the plant. *Determinate* plants come to maturity at specific times and don't put out any fruits or vegetables after that time. *Indeterminate* plants grow all season long, until the first fall frost kills them off. This means you can usually find a variety that's well-suited for your climate and the length of your growing season. You may even consider growing fruits and vegetables with varying growing seasons to ensure you have fresh foods to eat all summer long.

Really, the best thing to do to determine what grows well in your area is to ask at the local home and garden center. Or call the cooperative extension service and have a chat with a master gardener — he or she knows all there is to know about growing plants in your area.

Individual plants often have ratings that correspond to the USDA's Plant Hardiness Zone Map (www.usna.usda.gov/Hardzone), which categorizes regions by average temps and frost dates so you can figure out which plants are likely to grow well.

Resisting pests and pestilence

Modern technology is at work developing plant varieties that are resistant to all manner of diseases. You can find specific potatoes, eggplants, apples, tomatoes, and so on that allow you to reduce your pest-control efforts.

You can even, after a few years of trial and error, come up with a plant that's acclimated to your area. Save the seeds from plants that do especially well and plant them the next season. After a few seasons, you'll end up with a plant that's just right for your garden — it likes the soil and isn't bothered by the bugs.

Selecting for size, shape, and flavor

Fruits and veggies such as peppers or tomatoes come in all shapes, sizes, and even colors. Some are known for a high yield, and some come in sweet or hot or other flavors. The possibilities for what you decide on are endless!

Take your available space into consideration as well. Within all veggie categories, you can get a specific variety that stays put, called a _bush plant,_ or one that spreads out, sending out shoots that travel across the ground; it produces new veggies until killed off by a frost. You can get tomatoes that are a bush variety or traveling variety, bush or traveling beans, and so on. Therefore, how much space you have shouldn't limit what you want to grow, but it may make a difference in which varieties of plants you pick.

Choosing special varieties

You find multiple varieties of plants within a species. You can know exactly what to expect year after year if you plant heirlooms, or you can do some experimentation and come up with your own hybrids:

✔ **Heirlooms:** Heirloom plants breed true to form. Whatever the parent was (the plant you got the seed from), you'll get that same thing again and again. An example is the Brandywine tomato, bred, you may say, for its excellent flavor.

✔ **Hybrids:** These types of plants are a result of cross-breeding, and they're often quite hardy. Plants with specific traits were crossed to get a resulting plant with a specific combination of traits from the parent plants. If you save the seeds for next year, many of those plants will display a different combination of traits.

Deciding Where and When to Plant

Individual plants have specific needs. In this section, I discuss how plant type can affect the ideal location and timing of when you put your plants in the ground.

Checking factors for good growth

You want to make sure that you plant the right stuff in the right location. Plants that like full sun aren't happy if you plant them under a tree. Those that need a lot of water don't do well in an area with poor drainage. Consider the following growth factors:

- **Temperature:** Be sure to pay attention to frost dates, and don't plant until after the danger of frost is over. Daytime temperature sometimes matters. Arugula is one plant that doesn't do well in the heat of midsummer and should be planted in the spring or late summer.

- **Sun versus shade:** For most crops, the area should be shade-free. Veggies and fruits need a lot of sun. Other plants, such as some flowers or bushes, may do better in indirect light or even shade.

- **Amount of daylight:** Actual length of day may make a difference as well. Onions, for instance, need a certain number of hours of sunlight before they form a bulb. You need to choose the right varieties for your region and make sure you plant in an area that gets enough light.

- **Soil type and preparation:** A small area with beautiful, loamy soil can yield more crops than a larger area with not-so-beautiful soil. Soils that aren't conducive to healthy plant growth (for instance, sandy soil or clay) may have to be amended.

  Sometimes your soil just doesn't support plants, so if you're planning a small plot, you may consider constructing raised beds.

- **Drainage:** The space should have pretty good drainage. Most plants don't like to be constantly flooded. A lot of water runoff can form ruts that undermine the very soil the roots are trying to hold onto.

- **Water:** Your plants need to be in a place where large trees don't hog all the water.

- **Wind:** Strong winds can have a drying effect and can also rip plants apart.

Planting according to season

Growing plants means more than just putting seeds in the ground and then providing ample water. You have to get the timing right.

Looking at first and last frost dates

Frost kills. You can look at charts that explain when to expect last and first frosts of the season (the Victory Seed Company shows various cities within your state at www.victoryseeds.com/frost). These dates bookend the growing season so that you know when not to put plants outside and when the season is going to be over for good. These figures are estimates and give you an idea of when it's safe to plant. To be even safer, check with your local weather service.

Sowing seeds or transplanting larger plants

Certain plants need to be planted at certain times, and if the proposed time isn't a good time to plant in your area, you may have to go with an already started plant later in the season. If you have a greenhouse, start those plants from seeds yourself several weeks before you plan to put the plant in the ground (see Chapter 14 for details). If you don't have a greenhouse, acquiring plants from a neighbor or home and garden store at planting time is a good option for you.

Planting rounds two and three: Succession planting

Determinant plants have a specific growing season. To get the most out of your garden, plant some early-season determinant plants in the spring, and when that season is over, plant a warm weather plant. When that season is over, you may even be able to plant a fall crop. Table 13-1 shows how you can have one crop follow another.

As a means of protecting the soil over the winter, consider planting a cover crop in the fall. Grasses, legumes, and grains are all good choices. Then in the spring, turning them under adds nutrients to the soil in preparation for the planting season.

Table 13-1	Crop Succession Ideas	
Spring Crop	*Summer Crop*	*Fall Crop*
Peas (when they finish producing, turn the greens into the soil)	Corn	Cabbage or broccoli
Radishes, kohlrabi, turnip greens	Tomatoes and peppers	Beets, spinach, and chard
Lettuce	Squash	Broccoli

Rotating crops

Just as some people devour all the cherry cordials out of a box of chocolates while ignoring the coconut creams, some plants display clear nutrient preferences. Plant the same thing every year, and certain nutrients may become completely depleted from the soil.

The solution isn't simply a big dose of fertilizer. To keep the soil at its healthiest, you want to rotate plants. Plant crop *a* in the southwest corner of the garden one year and put crop *b* in that corner next year, moving crop *a* to a different corner. And don't plant just anything after crop *a;* the type of plant makes a difference to the health of the soil.

Plants that are particularly adept at sucking nutrients out of the soil should be followed with a plant that puts it back. For instance, peas and other legumes put nitrogen back into the soil and thus are good ones to plant after a crop that sucks nitrogen out, such as corn. Table 13-2 shows a possible rotation schedule.

Table 13-2	Crop Rotation Ideas	
First Year Crop	*Second Year Crop*	*Third Year Crop*
Peas	Corn	Cover crop (grasses, legumes, or grains)
Tomato	Radish	Cabbage
Green beans	Squash	Lettuce

Rotating crops has another purpose as well. If you keep planting the same thing in the same place, any diseases that like that plant can take hold in the soil. Plant diseases don't typically cross over to other plants, so you should instead plant something different in successive years. Crop rotation helps control plant diseases and insect pests by setting up a less welcoming place for bugs, bacteria, and other baddies to stay. The University of Washington Extension has a nice Web page (mtvernon.wsu.edu/plant_pathology/veg_rotations.htm) with a pretty large list of ways to use crop rotation to get rid of pests and diseases. This table lists the plant, the disease it's susceptible to, and the rotation suggestion to keep it from developing.

Succeeding with Vegetables

With so many options for veggies to pour your time, money, and heart into, the possibilities may seem endless. Some limitations that can help you make your decision are the amount of space you actually have to devote to crops, the quality of your soil, and the growing season, although certain varieties may be able to compensate for those shortcomings (see the earlier section "Garden Variety: Choosing Plant Types" for details). The ultimate decision may come down to what you like to eat.

Although what grows well may vary according to your soil and your skills, this section lists veggies that hobby farmers typically grow pretty well. Whatever you plant, and however many you decide on, you should probably plant at least two plants so that if something doesn't mesh with one, you have a backup plant. Some plants even require partners so they can pollinate each other.

Because your livelihood doesn't depend on having a successful crop, feel free to experiment with what and how much you plant. As you become more confident about both what grows well in your area and how much time and effort you can afford to invest, you can expand or contract your garden as needed.

Squashes

Although many varieties of squash are available, a couple consistently show up in gardens because of their ease of growing and their tastiness and familiarity to the consumer.

Squashes come in two basic varieties: summer and winter. The summer varieties, as the name suggests, grow well in the warmer months and are delicious eaten right away. Winter squashes favor the cooler months. But besides the growing season, they differ in another way. Summer squashes are those with a soft skin that you pick and eat all summer. Winter squashes have a hard shell and do better in storage than do the summer varieties.

Summer squashes: The prolific zucchini

Summer squashes thrive in warm weather and have a soft, pliable shell. Of the several varieties that abound in the summer squash class, the most popular is the zucchini. It's a prolific plant, and its fruit is versatile — you can eat it raw in salads, steam or grill it as a side dish, or even use it as an ingredient in a sweet, cinnamony bread! See Chapter 15 for a bread recipe.

Unless you really like zucchini, or unless you plan to sell the things and have a reliable recipient, don't plant more than a few plants. They'll yield dozens of veggies throughout a season. Their mission in life is to produce offspring, so if you keep picking the vegetables, the plants will keep trying to be fruitful and prosper. A friend of mine has stated that it's "against the law to plant more than two" because of how prolific this plant is. Really!

Bush zucchini also grows well in plant pots. Say you have a nice long deck that would benefit from some beautiful vegetation. Consider planting a zucchini plant or two in containers and placing them on that deck. Zucchinis are nice-looking plants, and if placed up on the deck and away from the soil (and the insect and the other wildlife pests), they may thrive even more than they do in the garden. Harvesting is even a little easier — just walk out on the deck every morning and pick off whatever veggie is ready!

Zucchini is one of those plants that pretty much grows itself. Toss two or three seeds directly into the soil (which you've mixed with fertilizer or compost) in late spring. Give them plenty of water and add organic fertilizer every four weeks or so. When the seedlings are about 4 inches tall, gently pinch off the weaker leaves.

Harvest the zucchini fruits when they're still small, maybe about 7 inches long (the size of a typical pickle). If you let them go, they continue to grow and can get downright huge — baseball bat-sized! Smaller fruits are more tender and contain fewer seeds. Of course, if a zucchini gets too big, you can grind it up for zucchini bread or consider feeding it to your critters — they'll love you for it.

Winter squashes

Many varieties of hard-shelled winter squash are available, and each type has its own amazing characteristics and tastes. Of all the possibilities, the most popular are pumpkins, butternut squash, and spaghetti squash.

Winter squashes are known for being easy to grow. Simply plant seeds in separate mounds about a yard apart and keep only the strongest seedlings — this practice keeps the plants strong and healthy.

After about 14 to 20 weeks, winter squashes are usually mature and ready to be harvested. Cut them off the vine, leaving at least 2 inches of stem. Store them in a dry place until you're ready to do something with them.

Legumes, the nitrogen-fixers

Legumes are plants that grow fruit in pods and help *fix* their own nitrogen — that is, they provide luxurious homes for beneficial bacteria that take nitrogen gas from the air and turn it into something the plants can use. In many cases, you don't have to fertilize the plants with nitrogen at all. These plants are also good to use in crop rotations because they return some nitrogen to the soil just by being planted there. The most popular legumes for small farmers are peas and beans.

Peas

Peas freeze and dry well, and with some peas, such as snap and sugar peas, you can even eat the pods. Peas are a cool-weather crop, so you should plant anywhere from late winter (provided your soil is workable) to early summer; then plant again in late summer/early fall. They don't do well in the heat of the summer.

To extend your harvest of fresh peas, try planting some in one area and then planting some in another area a couple of weeks later.

Because peas like a higher soil pH (lower acidity) than most vegetables, they benefit from having a little lime added to the soil along with compost or other fertilizer. (*Note:* The pH is a measure of the acidity or alkalinity of the soil. If the pH is a number below 7, it's considered acidic, and if the number is higher than 7, it's considered basic or alkaline. If the number is 7, the soil is neutral.)

Plant legume seeds every 2 inches or so, and give the plants something to climb up, such as small sticks or fancy metal or plastic poles from your local garden store.

Some varieties can grow and climb very tall, which means they produce shade and can interfere with other plants around them. A good plan is to use the space around the pea plants for smaller, shade-tolerant plants, such as radishes.

Soybeans

Soybeans grow quickly and are easy to deal with. Plant the seeds in warm soil after the last frost, and keep them moist. The richer the soil, the better, so add some compost. After two to three months, the pods are full-sized and start to flower, indicating they're ready to harvest.

You can eat soybeans pods and all, but to get them out of the pods, boil them for 20 minutes, let them cool, and then squeeze the pods — the beans will pop out. You can freeze or can the beans or put them right on your salads.

Green beans

Green beans come in bush varieties, which stand up on their own, and pole varieties, which need some sort of support to lean on. Plant pole varieties from seed about 4 inches–6 inches apart; bush varieties should be 2 inches–4 inches apart.

Harvest green beans regularly to encourage the growth of more beans. Pods should be about 3 inches long. Hold the stem with one hand while picking the pods off carefully with the other. Do the picking in the morning, when the beans are still tender but after the dew has dried off. Picking from wet plants can spread *bacterial blight,* a disease that threatens the bean plant.

As with peas, you can plant a bed of green beans every couple of weeks to extend the harvest throughout the summer.

Tomatoes and peppers: The not-so-deadly nightshades

The nightshades are staples in most home gardens, and their fruits are both tasty and healthy. Tomatoes and peppers are nightshades, as are eggplants and tobacco. (Potatoes are nightshades as well, but I discuss them later, in the "Root crops and tubers" section.)

Tomatoes aren't poisonous as people once believed, but the leaves and stems of plants in this category can be toxic. Take care not to feed the green parts to the animals.

Tomatoes

If you grow only one type of vegetable, tomatoes are a good choice — they're easy and yummy. In foods, they go with pretty much everything, and very few earthly delights can beat that. Tomatoes contain lycopene, which has been linked to prostate health and can help with some other forms of cancer and heart disease.

Tomatoes contain flavonoids (flavor pieces), some of which can be activated only with alcohol! If you cook tomatoes with white wine, you can unlock these things and get a better flavor.

Warning: After you've tasted a fresh-from-the-garden-tomato, you'll never be happy with a store-bought tomato again. Tomatoes you get in the store are picked before they're completely ripe so they'll last through the time it takes to get them from the farms to the shelves. These tomatoes do turn red, but turning red and ripening aren't the same thing. Vine-ripened tomatoes are the tastiest because as they ripen on the vine, they continue to get nutrients — and sugar — from the plant.

Bush tomatoes, the kind that don't ramble very far from the roots, set all the tomatoes at once. They have a short harvest season, but if you're canning, the bush tomato is a good kind to plant. Of course, they're not-so-good if you intend to eat fresh tomatoes all summer. Planting both varieties can ensure you have both fresh tomatoes all summer and canned tomatoes to enjoy in the winter.

Hundreds of varieties of tomatoes are available, from the itty bitty grape tomatoes to the massive beefsteak tomatoes, so-named because one slice can cover an entire sandwich. Plum or paste varieties such as the Roma are ideal for canning and sauces. Plant several types and find out which you like the best.

Start tomato seeds inside, six to eight weeks before the last killing frost. Starting them early gives them a chance to get big enough to be planted outside as soon as conditions allow. If you don't have a greenhouse where you can control temperatures, a good way to get the tomato plants started is to put them on top of your water heater. That gives them the warm and dark environment seeds like. Move them into the light as soon as they sprout.

Staking the plants has many benefits. It gets the vegetables off the ground, provides better air circulation, gives bugs fewer places to access the plant, and reduces the potential that a vegetable will rot on the ground. If you're going to stake the plants so they can climb, put the stakes in the ground first to protect the young roots. Put a little fertilizer in the soil as you plant the plants. An initial dose of fertilizer is enough, with maybe a little topical application later.

If you treat a tomato plant too well by giving it lots of water and fertilizer, you'll end up with a beautiful plant that bears little fruit. What another plant may consider a good amount of nitrogen, the tomato sees as something to beef up its green parts. The ideal tomato plant is a little scraggly with beautiful, healthy fruits.

Pick your tomatoes as soon as they ripen, and you can get fruits all summer. As soon as the first frost comes, the plants are done. (Anything left on the vine turns to slime.) If you still have some green tomatoes when a frost is threatening, pick them off the vine and put them in a window. They won't ripen anymore, but they'll get red and soft.

Peppers

Peppers come in a ton of varieties as well as a rainbow of colors. They can be hot or sweet, round or oblong, with colors including ivory, purple, bright red, yellow, orange, or even brown. Of course, although brown peppers taste fine, they may not be the most aesthetically pleasing things to put in salads.

All peppers are green peppers first. When they ripen, they turn into their other color. That means green peppers aren't technically all the way ripe. (Other plants work that way, too. Olives, for instance, start out green, and you can pick and eat green ones, but the ripe ones are black.) Green peppers have a specific flavor; when they ripen to their final color, their flavor also changes: They tend to get sweeter. Fully ripe peppers also have more vitamin C.

Peppers are related to tomatoes, so growing and caring for them is pretty much the same (see the preceding section). Too much fertilizer results in beautiful, healthy, green plants with little or no peppers. Prepare the bed, put in a little fertilizer and maybe compost, and put in the already-started plants.

Corn on the stalk

You can choose among three basic types of corn:

- ✔ **Sweet corn:** This is the stuff that goes to backyard barbeques. It doesn't store well and is best if eaten right away, because as soon as you pick it, the sugar enzymes start to turn into starch. The longer it sits, the less sweet it becomes. Plant two to three varieties a few weeks apart to extend the harvest.

- ✔ **Field corn:** This corn grows really tall (like 12 feet high) and has little or no sugar in it. Its purpose in life is to be grown, dried, and then ground into corn meal or used in animal feed.

- ✔ **Popcorn:** Yes, you can grow your own popcorn. If you have kids, it could be kind of fun. But for as cheap as you can get ready-to-pop corn, growing it may not be worthwhile.

Corn plants are considered a heavy feeder, meaning they really abuse the soil they grow in. They leach out any nutrients they can. Corn plants especially suck out a lot of nitrogen, so fertilize relatively often.

For best results, plant a block of corn instead of rows. With corn, the pollen has to fall from the tassel onto the ear shoot in order to produce a mature ear. With a block (kind of a square area, maybe four rows planted with corn for about 5 feet), the plants can more easily cross-pollinate each other. The plants in the middle of the block get the most benefits, yielding the better ears of corn.

Corn likes a lot of water, but not from above, which washes away the pollen needed to make the ears. The best watering method is a drip system. In its simplest form, you simply lay a hose along the rows, cut small holes in the hose, and turn the water on at a low pressure.

Leaf crops and cruciferous veggies

In leaf crops, the leaves are the tastiest and most nutritious parts. Also in this section is a subcategory called *cole crops,* or *crucifers,* vegetables that belong in the mustard family, such as cabbage, Brussels sprouts, broccoli, cauliflower, collards, kale, kohlrabi, turnips, and watercress. All are delicious, but this section simply focuses on some of the more common and easy ones.

Lettuce and spinach

Lettuce and spinach are so good, and the animals love them, too. The main types of lettuce are leaf lettuce, butterhead (bibb is a popular variety), Romaine (also known as *cos*), crisphead (such as iceberg), and stem (or asparagus) lettuce. Spinach comes in three main varieties: savoy (what's sold in most supermarkets), smooth-leaf (which has larger leaves than savoy and is typically grown for canning and freezing), and semi-savoy (a hybrid between the two).

Lettuce and spinach favor cooler temperatures, so plant them in late spring or early summer. They like small amounts of water, administered often. Cut the leaves whenever you think they're big enough. The plant has a shallow root system, so be careful when you harvest — don't tug too hard or the whole plant could come up.

Lettuce and spinach are easy to grow, but they grow in such a way that slugs like to hang out in the leaves. Some other common pests include aphids (watch for a gathering of these guys on the underside of the leaves) and foliage rot, caused by too much water and too-hot temperatures.

Cabbage

With the many varieties of cabbage around, you should really try growing at least one. Green cabbage is by far the most popular, but others such as red or savoy can also be a nutritious and tasty addition to your diet. Cabbage contains vitamin C, and all sorts of good stuff are in sauerkraut, which is simply fermented cabbage; it boosts the immune system, helps fight cancer, and helps your digestive system run smoothly.

Way back in the days of Captain James Cook (1728–1779) and the exploration of the seas, pickled cabbage was always in abundance on the journeys as a weapon against *scurvy*, or vitamin C deficiency. Cabbage preserved from the summer's harvest was also used in winter months as a vitamin C source when fresh fruits and vegetables weren't available.

Cabbage likes cooler temperatures, and you can grow it either in the spring (plant six weeks after the last frost) or fall (plant eight weeks before the first frost). The trick to getting good plants is rich soil, good fertilization, and plenty of water.

To help warm up the soil more quickly for the spring planting, lay black plastic on the ground and poke a few evenly spaced holes where you'll then put the seeds.

The appearance of a head of cabbage doesn't necessarily mean it's ready for harvest. Give it a little squeeze, and don't harvest until the head is nice and firm and dense. Cut it from the base of the plant.

Brussels sprouts

Brussels sprouts are sweet and tender and delicious when steamed. They're very similar to cabbages and can cause a not-so-pleasant fragrance when cooked.

These plants do well in warmer temperature, so you should plant them in early to mid summer. They do their growing during the heat of the summer, so be sure to keep them good and watered.

When the buds (or sprouts) get to be around 1 inch in diameter, you can pick or cut them off. As soon as the leaves start turning yellow, the plant is at the end of its growing season, and you should pick any remaining sprouts.

For best results, pinch the growth off the top of the plants after they're at least 2 feet high. This allows the plant to concentrate its energy on the buds in the lower part, making those buds denser.

Root crops and tubers

Root crops are those whose roots or underground bulb is the tasty thing you eat. Several plants fall into this category. The easiest to grow are the radish, onion, and carrot, but others include turnips, beets, parsnips, artichokes, leeks, garlic, and sweet potatoes. In this section, I discuss both root crops and *tubers,* which are enlarged underground stems.

Carrots

Carrots are a great source of vitamin A, which aids in eyesight, builds the immune system, and promotes bone growth. They're a cool-season crop, and they need a lot of water, so they probably won't do so well in a dry climate. The best temperatures for growing carrots are between 60 and 70 degrees. You sow carrot seeds directly into the soil.

Potatoes

Potatoes, like tomatoes, are a member of the nightshade family. These tubers can be eaten at various stages of maturity, all of which are tasty and healthy, from the young *new potatoes* to the fully mature.

Potatoes need a soil that isn't sopping wet, but they can stand cooler temperatures, so an early spring planting works for them. Start plants by planting *seed potatoes* — use only certified ones that you get from a nursery to ensure they're disease free. A couple of weeks before you intend to plant, set these special potatoes somewhere that's warm and well-lit. When planting day comes, cut the now-sprouting seed potatoes into small chunks, making sure each chunk has at least one eye, and let the chunks sit for a day or two.

Mix the soil with some compost and put the chunks in the ground, maybe 15 inches apart. Give the plant a lot of water while it's growing, stopping at the end of the season, when the foliage starts to turn yellow and die. Let the tubers sit and mature for a couple of weeks and then harvest.

Too much organic matter can be damaging to potato plants and can cause a disease called *scab,* which makes them aesthetically unappealing but doesn't ruin their nutritional benefits.

Radishes

Radishes are some of the easiest vegetables to start out with, and indeed, the radish was one of the veggies that came out of my very first garden. They do well in all climates and mature really quickly, going from seed to table in as little as four weeks. And because they grow so quickly, the bugs usually don't have a chance to infest them.

Mix compost into the soil prior to planting, and then all you have to do is provide sufficient water. Instead of planting in rows, spread seeds out in a small patch (1 or 2 square feet). Harvest as soon as they get to be a size that looks good. If they get too big, they become woody, and after the plant goes to seed, the radishes aren't worth eating.

Don't plant radishes where the dairy goats can get into them, and don't feed them to the goats. Radishes negatively affect the taste of the milk.

Onions

The first year you grow an onion, the plant forms a bulb. The second year, the plant goes to seed if you haven't harvested the bulb, so you can save the seeds but the bulb is no longer useful as a food product. Thus, the plant lives for only two seasons.

The length of the day determines when onions form a bulb, so certain types of onions work best for certain parts of the country:

- ✔ Long-day onions, such as the Walla Walla, are better suited to gardens in the northern half of the U.S.

- ✔ Day-neutral onions, such as the Stockton Red, do well anywhere.

- ✔ Short-day onions, such as the Vidalia, grow better in the southern part of the U.S.

If you're starting onions from seed, plant them in April so the plants have a chance to get strong before the warm weather and amount of daylight triggers the bulbs to begin forming. If you're using seedlings or sets (small bulbs), wait about another month so the colder weather doesn't stimulate seed development instead of bulb development.

When most of the tops have fallen over, the onions are ready for harvest. Pull the bulbs out of the ground, rinse them off, and let them dry out for a couple of days. Then cut the tops and roots and let them sit another day or two. Now they're ready to eat or store for later.

Leeks

Cousin to the onion, the leek is a little easier to digest and makes a great addition to many recipes where you'd otherwise use onions.

Leeks are a cool-weather crop, and they like full sun and well-drained soil. They need a reliable watering schedule — about 1 inch per week — or the stems get too tough. To get them to grow large, succulent, and white, blanch the lower part of the stem by piling dirt around the stalk as it develops. In midsummer, start removing the top half of the leaves to encourage stalk growth.

They're typically ready to harvest in the fall, when they're about as thick as a finger.

Stem crops

Asparagus, rhubarb, and celery are three types of plants whose edible parts are the stems. Other stem crops include hay and alfalfa, but because this section concerns vegetables and because celery isn't very easy to grow, the only ones I discuss here are asparagus and rhubarb.

Asparagus

Unlike most other vegetables, asparagus is a perennial plant, meaning it comes back year after year. In other words, you only have to plant it once. If you care for it properly, it may last 15 years or more.

You usually have to wait a couple of years before the plant bears any edible shoots. In the first and second seasons, the plant concentrates on building up a stable root system. Around the third season, you finally see spears that you can harvest.

Wait until the spears are about ⅜ of an inch or more in diameter before you harvest them. Make a cut just below the ground to enable the cut to properly heal, or snap off the spears at soil level. You can expect to get eight to ten sprouts per plant the first harvest year and more as the plant continues to mature.

Never take everything when you harvest asparagus; instead, plant extras and pick what you want without depleting the whole crop. Leaving some makes for a healthier crop.

Asparagus does especially well in moist environments. Plant them on banks of a canal, creek, or river you have running through your property and benefit from healthy plants that also help keep bank erosion down.

Rhubarb

Another hardy cool-weather plant, rhubarb is drought resistant and can do well in most soils, but it loves well-drained soils enhanced with organic matter.

To get the best stalk growth, remove flowers as they appear. Don't harvest the stalks until their third year. In the first two years, the stalks and leaves are nourishing the roots. Beginning the third year and ever after, the stalks may be picked all at once or individually over the next four to six weeks. After the stalks are cut, you can remove the leaves. (The leaves are toxic, so don't feed them to animals.)

Planting Other Small-Scale Crops for Food and Profit

Besides vegetables, other crops may do well on your farm and add some joy to your menus or those of your friends and neighbors. If you have an interest in making a little money on the side, you can grow these flowers, fruits, and nuts and market them to local restaurants or grocery stores. You may even get a booth at a farmer's market or set up your own roadside stand. (Be sure to check with local laws to be sure you're legally selling your wares.)

Letting business bloom with flower gardens

Who doesn't like the beauty of colorful and fragrant flowers adorning the landscape around the house? Besides adding to the aesthetics of your own outside look, you may consider growing flowers to either grace your own table (or desk or bureau top) or provide the materials for drying or pressing for subsequent use in crafts (see Chapter 17 for ideas). Or you can sell fresh flowers to the local flower shop.

Although the staples such as roses, carnations, and daffodils are popular and often in high demand in local floral shops, you may find an even more lucrative side business in growing exotics such as lilies or orchids.

Ask around. See what the local flower shops are lacking, and find out whether they're open to accepting product from a small farmer.

Aiming higher with fruit- and nut-bearing trees

Not everybody can grow fruit and nut trees, but if your soil supports them, they can be a lot of fun. Fruits and nuts are a great way to supplement your own diet as well as your income. The best way to figure out what you can grow is to check with a local nursery.

Fruit trees

How cool is it to go out into your backyard, pick a fruit, and then take a bite, tasting a tremendously sweet, healthy snack?

You can choose fruit trees in dwarf, semi-dwarf, or standard sizes. If you choose a fruit that requires cross-pollination, such as apples or plums, you need at least two trees.

You should prune, water, and fertilize your tree to keep it as healthy as you can so it produces the best (and most) fruit. Bees are a godsend for pollinating the flowers.

After you have a good, reliable supply of fruit, contact local grocery stores or even consider setting up a booth at the summer farmer's market or a roadside stand.

Nuts

Historically shunned by the average dieter for being high in calories and fat, nuts are welcomed with open arms by hikers and backpackers and others who regularly do some moderate to heavy exercise. However, people are reconsidering the health benefits of nuts for daily consumption. They're high in protein and rich in fiber, and they contain antioxidants. The fats they provide are the healthy ones — the much lauded omega-3s — so there's definitely a market for nuts.

Several varieties of nuts work on a small farm operation. Growing up, I was regularly treated to English walnuts from the big old tree in my backyard. Nuts grow better in loamy or sandy soils and don't do as well in soils with a lot of clay.

Prune, water, and fertilize your tree as needed. A new tree takes three to four years to be mature enough to produce nuts, and after the nuts start coming, they'll continue every year until the tree dies.

Another idea for using the nut tree is to sell the wood. A lot of nut tree wood is used in smoking. For instance, people often use hickory tree wood in smoking bacon.

The black walnut tree produces *juglone,* a substance that can be toxic to plants growing around the tree's base. Avoid planting sensitive plants such as tomatoes, peppers, cabbage, or eggplant nearby.

Growing berries and grapes: Fruit bushes and vines

Growing berries yourself means the fruits can be more natural — there'll be no pesticide residues or preservatives — but that also means they need to be eaten, dried, frozen, or baked within a few days of picking. Homegrown berries can be a lot cheaper than store-bought ones — you may even sell them to the store!

Raspberry bushes

Raspberries are versatile. Peach melba wouldn't be the same without them, and raspberries make a beautiful jam or preserve. And who doesn't like raspberry vinaigrette, adding a hint of sweetness to a salad?

These berries grow in the wild. Have you ever taken a hike — or even a walk in a city center — and seen wild raspberries growing by the side of the road or trail? Well, that goes to show you how easy they are to grow in your own yard. Only the nastiest of soils disallow a raspberry bush.

Raspberries are typically started from a crown: An established plant throws out underground shoots, and those shoots periodically throw a stem and leaf up out of the ground, a new plant called a *crown.* Be sure to get crowns from a place that can certify the crowns are disease free — most home and garden stores are a good bet. Then harvest as you see the berries appear. It takes a raspberry bush two to three years to produce berries.

Strawberry plants

Fresh strawberries are one of life's little treats. You can put them on waffles, on cold cereal, or into the blender to make strawberry or strawberry-banana milkshakes. You can make jam or even bake them into muffins. And don't forget just dipping your hand into a bowl and eating them au naturel!

Strawberries come in two basic types:

✔ **June-bearing:** These are usually the better-quality berries. They set all at once, giving you gobs of berries at one time. If your intention is to make jams, these are the ones to use.

✔ **Everbearing:** This type produces a few berries at a time, but in a more extended growing season. If you want to have berries as snacks or as additions to your desserts, these are the ones.

Strawberries, like raspberries, do better if started from an established plant. In a process called *daughtering,* a strawberry plant sends out above-ground shoots. Plant one of these daughters directly in the ground and wait for berries.

The way to keep the strawberry plant healthy is to periodically prune back the older or weaker, scrawnier parts of the plant, allowing the newer or stronger parts to thrive.

Don't plant strawberries where you've planted nightshades (tomatoes, peppers, potatoes, or eggplants). Nightshades can harbor *verticillium wilt,* a serious strawberry disease.

Grapevines

Grapes are a wonderful fruit. You can eat them fresh, dry and preserve them as raisins for snacks or in cooking, or mash them up and ferment them into a fine yet homegrown wine. My dad made wine when I was a kid, and he found it so rewarding. He had a secret recipe, and it was very sweet (of course, my brothers, as boys often do, broke the code and figured it out).

Grapes are planted from a starter plant. As the plant grows, you need to train the plant to grow upward (or overward). Install a trellis so the plant can travel in the right direction.

You don't get any fruit until the second year. The first year, the plant develops the stem, which turns into wood. Afterward, the wood, dormant in winter, produces new leaves and fruit each spring.

Growing Herbs for and in the Kitchen

If you decide you don't have the time to invest in a garden or a larger crop but still want to grow something, you can always plant some herbs in small pots and keep them indoors. If you have a few well-lit windows, these plants can supply you with exotic flavorings year-round. I list some of the easiest to get in Table 13-3, but there are tons of others.

Indoor herbs do better if they're well ventilated and away from cooking fumes. Typically, the bushy, perennial herbs — rosemary, thyme, oregano, sage, and winter savory — perform better indoors than those with soft stems, such as mint and tarragon.

Table 13-3	Types of Herbs and Their Uses
Herb	*Uses*
Anise	Has the scent and flavor of licorice; very good in salads and baking
Basil	Probably the most popular homegrown herb; used in so many foods, especially pesto and tomato-based recipes
Caraway	Use it to flavor rye bread; it's also good in coleslaw and Hungarian-style dishes and meat stews
Catnip	Cats go into a hypnotic trance when they get a whiff of catnip; it's like marijuana for cats, although it's a harmless treat for them
Chives	An onion-like plant that's popular with baked potatoes
Cilantro	Use cilantro in foods such as chili, salsa, and pico de gallo (Note: The seed, coriander, is common in Indian curries; ground, it's in many baked goods)
Dill	Seriously yummy with potatoes (consider sour cream and dill dip with potato chips, in potato salad, or on baked potatoes), in sauerkraut, or with pork roast
Fennel	Has an anise-like flavor that becomes delicate when cooked; use it in salads, in cheese spreads, or on vegetables or fish
Lavender	Dry and use it in sachets and perfumes; the fragrance is amazing
Marjoram	Use it in perfume; also good in cooking — sprinkle it over sliced tomatoes or baked fish
Oregano	Adds that extra spark to Italian-style recipes, especially pizza, lasagne, or good old-fashioned spaghetti
Parsley	High in vitamins A and C, this herb is versatile; use it as a garnish, to flavor soups, meats, or pastas or in ethnic foods such as tabbouleh or parsley pesto
Mint	Besides the food-related uses of mint (tea is one of the biggies), mint's a good plant to grow and place strategically to ward off mice
Rosemary	Seriously yummy when you rub a pork roast with rosemary before cooking
Sage	Add sage to stuffings or sprinkle it on most meats, especially poultry, rabbit, pork, and baked fish
Savory	Comes in a summer and a winter variety (summer is sweeter); both are used to flavor meats and vegetables
Tarragon	Similar to anise; use it in salads, sauces, and marinades
Thyme	Especially good in slow-cooking beef dishes, clam chowder, or poultry stuffings

Get seeds or starts at a local nursery and ask the salesperson for suggestions. He or she may suggest something you've never heard of but that turns out to be just right for your favorite recipe. Use your herbs fresh or dry them for later use.

Scented geraniums make wonderful, fragrant herbal roommates, as do lemon verbena, basil, coriander, and some varieties of lavender.

Plant dill, fennel, and parsley, and you may end up with caterpillars from the black swallowtail butterfly. You may want to plant extra so you can transfer all the caterpillars to a single plant and still have some herbs for yourself.

Chapter 14

Getting Those Plants Started

Growing your own food usually means at least some tender, loving care on your part. Weeds (those plants you aren't interested in allowing to propagate on your grounds) just grow by themselves with no help from you, but typically, coaxing the things you do want to grow takes a little effort.

In this chapter, you discover where to plant, how to prepare your soil, and what technology is available to help you get your crops started and keep them growing strong and healthy. You can even start plants in a greenhouse and then put them in the ground outside when they're well on their way. This chapter also addresses organic gardening. Get ready to dig in!

Planning Out Your Planting Space

You can't just decide to grow something, put it in the ground, and expect good results. In many cases, what grows for you depends on where on your property you put your garden and even on how you position the individual plants. Here are a few ideas:

✔ Choose a place that gives you easy access to your plants.

✔ If you're starting with a smaller garden, place it in an area where you can expand the garden in the future. If you're going to plant acres, consider setting aside some space so you can rotate crops (see Chapter 13 for details).

✔ Look for a relatively level area. You don't want all the rainwater on your property to pour into your garden after a heavy rain.

- ✔ Check water availability. Will the hoses reach? Or how far do you want to dig for underground systems?

- ✔ Think about shade. Most crops need as much sun as they can get. Planting right next to your house means plants are only in the sun for half the day and in shade the other half.

  Put shorter plants on the south side of the garden and taller plants on the north so the tall plants don't block the sun.

- ✔ Some plants like to run (they send out shoots and travel away from the root system), so give them a fence to run up so they take up less space.

- ✔ Garden design can be an issue. Do you want to plant rows or use square-foot gardening, in which you plant in a grid instead? Corn actually does better in a grid because each plant helps pollinate those around it.

Wherever you locate your garden, you probably want to limit access to the garden with fencing; otherwise critters can get in there and destroy the plants. Deer eat everything. Chickens scratch, eat plants, and build wallows where they dust themselves off. Dogs, too, wander in and lie down on the cool green plants, squashing the fruits and breaking stems.

Making Sure You Have Good Soil

Fertile soil is a mixture of water, air, minerals, and organic matter. The ratios of each element in the mixture vary widely, depending on your location and what you do to the soil yourself. Some soils may not have the right nutrients to allow any sort of wanted growth. (Weeds, however, usually grow no matter how bad the soil is!) Treat your soil well, and your soil will treat your plants well. This section explains good soil care.

Having your soil tested

Typically, you need to help the soil along in some form or other. Get a soil test from your local cooperative extension service to give you an idea of what kinds of nutrients you need to help your soil.

Typically for less than $10, someone from the extension service can analyze your soil and tell you what's deficient or overabundant and needs to be curtailed. The tester checks the soil's pH (acidity) level, its nutrient content, and the amount of organic matter that's in it. You also receive advice on what to do to correct the problems as well as tips for the specific plant types you want to grow.

Tilling (Or otherwise digging up the soil)

One of the first things most gardening books have you do is somehow work the soil. Why? Here's what tilling does:

- It aerates the soil. Soil can get compacted due to pets and foot traffic or vehicle traffic. Roots need air to thrive, so loosening the soil up helps get that air back in. Soils with a lot of clay compact more easily than those that are sandy or rich with organic matter.

- Tilling breaks up clumps that can interfere with root movement and growth.

- It blends in some sort of organic matter or nutrients your soil may be lacking. (I go into more detail about this in later sections.)

A friend of mine swears by what he calls "double digging" to move the soil around. He digs two long trenches about 1 foot wide, 1 foot deep, and the length of the planting bed. He then moves the soil to the opposite trench, working in whatever nutrients he is using.

Till near the surface (no deeper than 12 inches), or you'll be digging up more-compacted soil and mixing it in, making it more difficult for roots to spread out. (See Chapter 6 for some of the tools you need for working your soil.)

Minerals for vegetables: Fertilizing

Sometimes your soil is deficient in certain ingredients necessary for healthy plant growth. So just like all those vitamins and supplements you take yourself, the soil needs some supplements.

Using organic fertilizer from plant sources

If you stop to think about what *organic matter* really is (plant and animal material that's in the process of decomposing), you may be reluctant to use it around plants you eventually want to eat. But in reality, working this stuff into your soil is good for the soil and ultimately for the plants you intend to grow in it.

Fully decomposed organic matter is called *humus,* and it's important for soil structure. (For info on creating humus through composting, see "Composting 101," later in this chapter.) A good structure helps the soil

- Retain nutrients
- Allow oxygen and carbon dioxide to move freely through the soil
- Properly drain water

Undercover! Plowing under cover crops

Some growers raise a cover crop and then turn the soil under, allowing the plants to rot and become a natural fertilizer. A *cover crop* is something you grow specifically for soil enhancement. Choose anything that has nutrients and grows quickly, such as legumes, peas, vetch, or buckwheat.

Buckwheat has an added benefit of fighting weeds. One idea is to plant four to five successions of buckwheat in the same spot during the regular growing season for your first year,

turning it under as soon as it's done growing and before it goes to seed (around four to six weeks). That has the effect of killing off most of the weeds in that area while also improving the fertility of the soil. The second year, start your garden.

One problem you can run into with cover crops is that if you wait until after they go to seed to turn them under, you introduce those seeds to the soil.

Using organic matter in your soil is a way of creating a sustainable agricultural environment, or *permaculture*. For everything you take out of the ground, you put something back in. Doing this gives you benefits from your land (the fruits and veggies that you take away) without depleting natural resources. Plants don't care *where* nutrients come from, and if those nutrients can come from sustainable sources, all the better for the environment.

Alfalfa is almost pure nitrogen, so you can turn under any alfalfa your animals don't need, incorporating it into your soil. If you grow corn, after harvesting the edible parts of the plant, you can then chop the remaining parts of the plant and mulch it into your soil. A chipper or shredder can do the trick if you want to spread the parts in other places, or you can use a tiller right in the field.

Remember — working some organic matter into your soil doesn't mean you're a full-fledged organic farm! See the later section entitled "Going Natural with Organic Farming" for more information on going organic.

Fertilizing with manure

You can go to your local home and garden store and find bags of manure, or as you drive along the country roads, you may see signs advertising manure for sale. (Man, the things you now think are normal to buy now that you have a new life away from the city!) If you have your own animals, you can use your own manures, but be careful — not all manures are equally usable.

Manures from animals that eat hay or alfalfa are completely usable (as opposed to manures from carnivores, which aren't a good choice). Goat berries or llama pellets, for instance, can be shoveled up and applied directly to your crops.

Don't use cow and horse manures without first composting them. Viable oat seeds will undoubtedly still be in the horse manure, and unless you compost it, you'll end up introducing weeds along with your soil aid.

Using chemical fertilizer

Chemical fertilizers come in many forms and many formulas. Three numbers are on the bag, and those numbers correspond to the amounts of three major plant nutrients — nitrogen, phosphorus, and potassium (NPK) — in the mix. Figure 14-1 shows a standard fertilizer label.

A good way to remember what NPK stands for is that the first number is *up,* the second is *down,* and the third is *all around:*

- **N:** Nitrogen, the first number, helps the top growth, or what the plant sends up.

- **P:** The second number is for the amount of phosphorus in the mix, and it deals with root growth, or what the plant sends down.

- **K:** The final number represents potassium, which deals with the general health of the plant or the all-around health.

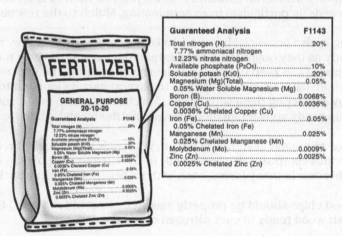

Figure 14-1:
The large numbers on the fertilizer label tell you the nitrogen, phosphorus, and potassium content.

Guaranteed Analysis	F1143
Total nitrogen (N)....................................20%	
7.77% ammoniacal nitrogen	
12.23% nitrate nitrogen	
Available phosphate (P_2O_5)...........................10%	
Soluble potash (K_2O)...................................20%	
Magnesium (Mg)(Total)...............................0.05%	
0.05% Water Soluble Magnesium (Mg)	
Boron (B)...0.0068%	
Copper (Cu)...0.0036%	
0.0036% Chelated Copper (Cu)	
Iron (Fe)..0.05%	
0.05% Chelated Iron (Fe)	
Manganese (Mn)...................................0.025%	
0.025% Chelated Manganese (Mn)	
Molybdenum (Mo).................................0.0009%	
Zinc (Zn)..0.0025%	
0.0025% Chelated Zinc (Zn)	

Fertilizer mixes can be *balanced* (each element has the same number) or *complete* (each number is represented, but not necessarily with the same number). Which one you need depends on your soil, what you're planting, and even the time of year.

Be careful with any chemicals you use on your land and attentive to what may happen if it rains a lot or if you simply forget to turn off your sprinklers. Follow the instructions on the bag, and apply at the recommended rate. Otherwise, you can face the following problems:

> ✔ Nitrogen is water soluble and can leach out of the soil. Overwatering drains this nutrient away from the plants you think you're helping, and you may need to reapply nitrogen. Also, an abundance of nitrogen in surrounding ponds or other water sources can cause unwanted algae growth. Excessive nitrogen can also burn and dry out leaves, or it can encourage leafy growth to the detriment of tomatoes and similar fruits.

> ✔ Soil particles containing phosphorus can pollute water sources. Remember when the powers that be took phosphate out of detergents because they were polluting the nation's waterways? Phosphate in water is not a good thing. Too much isn't a good for your garden, either, so don't overdo the application.

Adding a little sunblock and insulation: Much ado about mulch

Mulch blocks sunlight that would otherwise get to the soil. At ground level, blocking the sun is often a great idea. Sometimes you want to keep moisture from escaping your soil into the air. Or maybe you want to keep the ground cooler (mulch acts as an insulator). Or maybe you want to keep certain seeds (weeds, in particular) from germinating. Mulch to the rescue!

Mulching is also beneficial in winter to keep plants healthy if they're not as cold-hardy as they ought to be or to extend the harvest. Heavily mulching around plants can keep them alive longer by preventing the freeze-thaw cycle. For instance, you can continue to harvest carrots longer into the winter by protecting them with mulch. Mulches also help break down clay soils, help sandy soils retain moisture and nutrients, and minimize dust.

If you use an organic material as mulch, your mulch eventually breaks down and provides compost (also beneficial) to the surrounding plants. Mulching materials come in many forms, each with its benefits and drawbacks:

> ✔ Wood chips should be properly aged (composted for a year) because fresh wood tends to suck nitrogen out of the soil.

> ✔ Grass clippings are less likely to blow away but can be easily compacted.

> ✔ Pine needles are especially good on acid-loving plants, and they look pretty good, but they can be a fire hazard.

> ✔ Plastic holds in moisture well, but it can tear.

> ✔ Shredded paper is readily available but can blow away in the wind.

Don't confuse mulch with compost. Compost is a soil enhancer, put there to add nutrients; mulch is something you put on top to protect the soil from the elements.

Composting 101

Whatever is wrong with your garden, adding a little compost is likely to fix it. Yes, compost is magic, it's natural, and you can make your own out of left-over scraps. If your soil is too sandy, compost can help that soil hold moisture. If there's too much clay, compost can get in between the clay particles and loosen it up, allowing for better drainage and aeration. In this section, you get the scoop on compost.

Setting up a bacterial smorgasbord

Growing plants can't get nutrients directly from dead leaves and veggie scraps because the nutrients are trapped in cells. Luckily, a whole crew of bacteria (along with some fungi) is ready to help out. These bacteria consume the chopped up plant matter and other composting materials and break them down into simpler forms of nitrogen. When the bacteria die, they release the nitrogen, which becomes fair game for the plants.

Your job is to set up a compost banquet and create a welcoming environment for these beneficial bugs. In the simplest setup, you create a pile of organic matter (leaves and stems from the veggies you had for dinner, leaves and twigs from your trees), keep it moist, and allow it to cook. If you maintain the pile properly, you can get compost in as little as two weeks. In this section, I go over some of the elements of a good compost pile.

Perfecting the menu with a good carbon-to-nitrogen ratio

Optimal compost has a good carbon-to-nitrogen ratio, about 30:1. Carbon is especially high in brown, dead matter, such as sticks, straw, and fall leaves; more nitrogen comes from green things, such as veggie scraps and grass clippings.

So should your compost mix be mostly a pile of sticks? Not exactly. Ratios vary, but your veggie scraps may have a ratio of 15:1, and grass and cow manure may have a ratio of 20:1. Straw and leaves may have a ratio of 70:1. Therefore, your pile should be mostly green stuff, with a bit of brown, dry stuff mixed in. See the later section "Knowing what you can compost" for possible materials.

Letting the party heat up

As bacteria break down your compost materials, the temperature starts to rise. You want the compost pile to get to a temperature that allows "cooking" (around 140 degrees). If it goes above 140 degrees, life doesn't last very long, and your nitrogen-extracting bacteria begin to die off too fast. At 140 degrees, you have a good mixture of air, moisture, nitrogen, and bacteria working at optimal levels.

Get a temperature probe (it looks a lot like a meat thermometer with 3-foot stem to let you get into the center of the compost pile) to see how well your compost pile is cooking. If it's not up near 140 degrees, you need to do something to improve carbon-nitrogen mix, such as add some more twigs. The hotter the pile cooks, the more quickly the compost breaks down. If it gets too cold, you get rot and slime. (It'll eventually compost, but the most efficient way to compost is hot and fast.)

After your compost is done cooking, it's done for good (it won't compost any further), and you end up with humus — a material that's full of good bacteria. You can then store it. Put the compost into plastic bags to protect it from the elements.

Providing drinks

To understand the correct moisture for your compost pile, think about a sponge. When you soak the sponge and then wring it out so it's no longer sopping but is still damp, that's how wet you want the materials in your compost pile to be. If you get too much water in it, it smells like rotten eggs and the mass doesn't cook.

One of the standard fertilizers used in gardening is urea, a component of urine. I'm not saying you should go out and pee on your compost pile, but that would actually help it! Unless you have an active bladder infection, urine is sterile. Don't let your neighbors see you, though!

Keeping the air circulating with the guests

You need to turn the pile regularly and fluff it. Stick a probe of some sort in the pile — perhaps use a pitchfork or go as fancy as getting a turner (it looks like a 55-gallon drum on a center spindle). Lift and move the probe around to help with the air circulation. If you don't turn the compost, it'll start to cool down and go *anaerobic* (it'll decay without oxygen) and rot (and not smell so good).

Evaluating odor

Composting is nature's way of revitalizing the soil. That earthy aroma you encounter in a forest after a rain comes from the fallen leaves composting and returning plant matter back into the soil. That nice smell is what compost *should* smell like. If your compost pile doesn't smell good, something's wrong:

- ✔ If an ammonia smell is present, the mixture is leaking nitrogen (a major nutrient for plants and the first number in fertilizer). This can happen if you don't turn the pile often enough or if the carbon-nitrogen ratio is off. Try adding more sticks or other carbon-dry stuff.

- ✔ If there's no smell at all, the mix is too dry and you need to give it more water.

- ✔ If you detect a rotten egg smell, the pile is too wet.

To get well on your way to becoming a composting guru, check out *Organic Gardening For Dummies* (Wiley).

Knowing what you can compost

Technically, you can compost anything organic, but there are some limits. Here are some possibilities:

- ✔ Kitchen scraps (eggshells, old bread, and vegetable and fruit scraps) or vegetables past their prime

- ✔ Animal manure from herbivores

- ✔ Deadheaded flowers

- ✔ Trimmings (branches, twigs, and leaves) off of trees

- ✔ Hay and straw

- ✔ Sawdust

- ✔ Wood chips

- ✔ Shredded black and white newspaper (avoid colored ink)

- ✔ Although this sounds a little icky, you *can* compost a dead animal; pack it in sawdust, and after a couple of years, it'll be reduced to nice, helpful compost (see Chapter 18)

You can also compost diseased plants and weeds if the pile is running hot enough (around 140 degrees) and cooking really well; the high temperatures kill the pathogens and weed seeds. If the pile isn't hot enough, leave these things out.

Avoid using the following for compost:

- ✔ **Meats and fats:** These attract vermin and cause your pile to stink. Also, meats and fats make it harder to get air into the mixture.

- ✔ **Manure from dogs, cats, pigs, or humans:** Herbivore manure is okay; carnivore manure is not because it can introduce *pathogens,* disease-causing microorganisms.

- ✔ **Walnut leaves:** Walnut leaves contain juglone, a chemical that's a natural *preemergence* — that is, it stops seeds from sprouting. If you look on the ground around walnut trees, you may notice very few things growing.

- ✔ **Pine needles (unless you have a soil that's very alkaline):** Pine needles (or the whole Christmas tree for that matter) are extremely acidic.

Also, be careful with grass clippings. You don't want to add grass in the large sheets that form when piles of clippings matt together; sheets inhibit air circulation. Lay these sheets on your driveway and let them dry out before putting them in the pile.

Creating leaf mold, pure composting gold

Have you ever taken a walk in a forest after a rain? Can you remember the fragrance? Ah, I can smell it now. It's such an earthy, clean, good smell. What you were smelling (and what I'm remembering smelling) was actually leaf mold. *Leaf mold* is rotted leaves — or what's left of leaves after weather and time have worked their magic on them. It's a clean, green solution that contains no added chemicals, and it's totally organic.

Leaf mold improves soil structure and helps it hold water and nutrients. And it's easier to make than compost, but you use it the same way. Leaf mold is wonderful to add to alkaline soils because it's more acidic and can thus provide a balance.

You typically need to make your own leaf mold because it's not readily available commercially (perhaps because it's so easy to make — and who would really want to buy a bag of rotted leaves?). If you have lots of trees on your grounds, you don't have to do anything — just collect the stuff when it's ready. Without your own trees, arrange to pick up the leaves from a neighbor, toss them on your ground, and then when the leaf mold is ready, spread it around.

Here's an easy leaf mold recipe:

1. **Pile up the leaves and thoroughly wet them (either hope for rain or douse them with a hose).**

   Before watering the leaves, let the chickens have at it. They'll remove any unwanted bugs, resulting in purer compost.

2. **Cover the pile with tarps or plastic and toss a few rocks on top so the wind doesn't take the covering away.**

   You may want to put the leaves into plastic bags to keep them as moist as possible. Before closing off the bag, make sure they're wet and then poke a few holes in the bags with a pitchfork to improve air circulation.

3. **Wait six to twelve months.**

   The leaf mold takes a while to break down. Small leaves, such as alder, birch, or Japanese maple, break down in about six months.

   You may notice that worms have moved in. Where'd they come from? That's one of life's mysteries, but don't despair! Worms are considered nature's little composters and are amazingly beneficial critters to have around.

On the other hand, you can use whole, pre-composted leaves with your plants because putting them on the soil eventually results in a composted product that gives back to the soil.

Barring Intruders: Getting Weeds and Pests Out of Your Garden

Plants take a little care in order for them to be the best they can be. Your job is to make sure the weeds don't choke them out, the pests don't chew on them, and they get the right amount of water (see Chapter 8 for info on watering systems). This section presents some of the basics of weed and pest control.

Managing weeds

Most weeds just invade, grow, grow tall, and choke out the flowers and veggies you're deliberately trying to grow. You can take care of weeds in many ways. Not all weeds respond to all methods, so sometimes a combination is in order. Here are some options:

- ✔ **Mechanical removal:** Mowing, burying, tilling, and digging weeds up by the roots are mechanical ways of getting rid of weeds. These methods don't work too well on weeds that have deep root systems, such as bindweed (morning glory).

- ✔ **Chemical control:** Chemical pesticides can be very effective. However, you're venturing into the not-so-natural side of farming, and some people don't buy your stuff if you use chemicals, so chemical control isn't always the best choice in the long run.

- ✔ **Crop rotation:** If you continually grow the same things in the same areas and use the same herbicides, you end up with weeds that are tolerant to those surroundings. Rotating crops, getting different nutrients into the soil, and using different herbicides may confuse the weeds into submission.

- ✔ **Biological control:** This method uses living beings, such as beneficial insects or even livestock, to keep the weeds down. It works best in pasture areas.

- ✔ **Fire:** Periodically burning your fields, such as after each harvest and before planting, can keep some weeds down.

One person's weed is another person's wildflower. Although most of the invasive plants on my property are indeed considered something that needs to be eradicated, one of the crafts I dabble in is pressed flowers, and so many of those flowering weeds have turned out to be perfect supplies for my art projects. I even got mad at Hubby one time when he dug up one of the offending weeds that was messing up his garden. (The flowers it produced were so delicate and gorgeous and pressed so beautifully!) See Chapter 17 for some crafty ideas.

Guarding against wildlife and insect pests

A big part of gardening is trying to make sure you're the one who actually gets to enjoy the harvest, and that means keeping animal and insect pests out. Whether you want to stay all natural or go with the chemicals is a personal decision. Chemicals can be a godsend because they're fast and effective, but if you want your farm to be organic, you can't use commercial chemicals at all. Or perhaps you just want to go back to the roots of farming and try to do things the old-fashioned way, keeping your fruits and veggies healthy things to put into your body. In this section, I briefly explain some pest control.

Insects

Bugs can attack crops and destroy all your hard work. That's not fair! You know bugs have been at your plants if you see holes in the leaves or the fruit or veggie itself, but sometimes, as in the case of the squash bug, you don't know of the problem until you see that the plant is dying for no apparent reason. Squash bugs suck the juices out of the squash plant and inject a toxin that, along with the dehydration effect, adds to its demise.

Here are some precautions you can take against invading insects:

✔ Grow pest-resistant plants. Check with your local cooperative extension service for a list of what works in your area.

✔ Water the ground and not the leaves themselves because wet leaves are more apt to attract bugs (as well as make them susceptible to fungal diseases).

✔ Regularly remove any dead or dying leaves because these dead things provide a great place for bugs to hang out.

✔ You can get synthetic chemical pesticides or organic kinds. The chemical ones are poison, so be careful with their administration. Again, the cooperative extension service is the best place to get advice on what works best for you.

✔ Try some companion planting. Put plants that aren't susceptible to a certain insect infestation next to ones that are, and plant strong-smelling herbs next to vulnerable plants to help mask their scent. Include some plants that attract beneficial insects. For instance, dill and coriander attract good, tiny parasitic wasps and flies that eat various bugs that feast on plants.

✔ If you're dedicated to planning and management, try *biological control,* or using natural predators or parasites to kill specific unwanted bugs. Visit the Biological Control Information Center (`cipm.ncsu.edu/ent/biocontrol`) for more information.

Mice and other rodents

Rodents (mice, voles, moles, and prairie dogs) can wreak havoc on your property. They tunnel underground and leave behind unsightly runways through your lawn. And besides the lawn aesthetics problem, they also eat the roots of your plants or eat the fruits and veggies right off the stems. Here are some ways to keep the rodent population down:

✔ Get a humane trap that catches the critters live so they can be reintroduced into somebody else's neighborhood.

✔ Use the spring trap, which snaps a leg or neck. If rodents survive that injury, they're then rendered disabled and are less likely to be able to fend for themselves in the wild.

✔ Use fine-mesh chicken wire to build a cage around the root ball of a plant before putting it into the ground. The fine mesh keeps the rodents out while the roots are growing but is be wide enough to allow the roots to grow and eventually spread out of the cage.

✔ Kill the rodents with poison (though that's dangerous because other animals can get into it).

✔ Use the good old-fashioned barn cat. Or believe it or not, chickens are actually good mousers!

Birds

Birds can eat new growth from your plants, dig and scratch and get at seeds, eat entire seedlings, or sample the fruits and veggies after they mature. Cherries, peas, and corn are particularly attractive to birds. But don't despair — you can use a number of deterrents:

✔ **Physical barriers:** To protect fruits on trees, put netting over the tree.

✔ **Visual devices:** Visual deterrents such as the big old plastic owl or the scarecrow aren't very effective, mostly because they never move. (We put a big plastic owl on a deck post, and one day I saw a starling sitting on its head!) After a while, the birds come to know that these devices really aren't any threat. Put something out that moves, such as iridescent bird-deterrent foil.

✔ **Sound deterrents:** Some contraptions on the market emit a series of sounds when birds come near.

✔ **Chemical aversions:** If the birds are eating your fruit or vegetables right off the trees or vines, a biodegradable, kind-to-the-environment substance called methyl anthranilate is just the ticket. You spray it on your plants, and it renders them unpalatable to the birds.

Keep in mind that the right kinds of birds can help you eliminate your bug problems because birds eat bugs.

Deer (And other larger critters)

Although deer look so peaceful and docile (and I still love to see them around), they can be a real nuisance around the farm. They chomp on your newly sprouting veggies or chew on your trees. But fortunately, you can take some steps to live with them peaceably:

✔ Contact your cooperative extension service for a list of deer-repellent plants that grow well in your area. These plants can act as natural barriers to those other tasty-to-the-deer plants.

✔ Deer don't like the hair of other mammals. I've heard of putting hair (your own or your dog's) into a sock and hanging the sock on your fence to keep the deer from jumping over and getting to whatever's inside.

✔ Install a tall fence. You'd be surprised how high (and effortlessly) deer can jump. A 6-foot fence doesn't cut it — make it at least 8 feet high.

✔ To protect young trees, put them inside a mini fence. Make it tall enough to protect the tree and lower branches and set up the fence in a 1- to 2-foot diameter circle around the tree. The deer won't be able get their noses through the fencing and chomp on the young bark or buds.

You don't want rabbits, deer, or goats to get into your crops, so fences are a must. They can be elaborate, from sturdy structures that don't allow anything to get through (such as solid vinyl fences), to the gorgeous old-timey split rail fence, to a fence that incorporates electric shock wires, to the very basic metal pole with wire strung across. On my farm, we've put up T-posts that serve as an anchor for 2" x 4" wire mesh. (You can also use chicken wire, barbed wire, or single strand wire — the possibilities are almost endless.) Regardless of what you decide on, it has to be sturdy enough that your animals can't knock it over.

Going Natural with Organic Farming

Going *organic* means using only natural materials that once came from living organisms in your farm operations. That means no chemical pesticides or synthetic fertilizers of any kind. Instead, you use manures, composts, cover crops, and crop rotation to enhance the soil and use biological control methods (such as beneficial bugs) to manage pests. Sometimes that's a good thing (you aren't introducing anything into the environment that isn't natural), but it may prove to be a headache (you just can't seem to win the war with the bugs without chemicals).

Manures, leaf mold, and compost (all discussed earlier in this chapter) are easy organic additions to your soil preparation. For fertilizer, you can use a natural phosphorus source called *rock phosphate,* a slow-release 2–3 percent phosphate mix that you apply around every five years.

Here are a few types of organic pest control that you can use:

- ✔ **Diatomaceous earth:** This nontoxic insecticide is made of the fossilized, crushed shells of diatoms (kind of a hard-shelled algae). It feels like talcum powder, but to bugs it's like walking over crushed glass. It cuts their feet or bloats their stomachs.

- ✔ **Bacyllis tharangestis (BT):** BT is a type of naturally occurring soil bacteria. When ingested by bugs, the pathogen sets up home inside their bodies and ends up rotting their guts. This stuff is especially effective against certain families of moths, butterflies, flies, mosquitoes, and beetles. It doesn't affect humans, beneficial bugs, pollinators, or other wildlife.

- ✔ **Insecticidal soap:** This nontoxic soap contains fatty acids that clog breathing holes and soften up bugs so they dehydrate and die.

- ✔ **Other natural products:** Other organic products for staving off bugs that threaten your plants (and your potential food supply) include cayenne pepper sprinkled on an infested area, iron phosphate, horticultural oils, plant extracts, neem oil, and so on.

- ✔ **Pest-tolerant plants and mulches:** Bugs don't set up house in areas that contain pest-tolerant plants, so surround your plants with naturally pest-tolerant plants and use mulch around plants to create a bug barrier.

If you want to know more about going organic and doing it right, check *Organic Gardening For Dummies* or visit the Organic Trade Association's Web site on going organic (www.howtogoorganic.com).

Getting Inside Info on Greenhouses

Having a greenhouse is a great way to extend the growing season, which is really important in areas where winters are long and summers are short. You can grow plants inside the controlled atmosphere of the greenhouse during times when there's no way they'd grow outside. You can also start seeds in a greenhouse so the plants are already hardy when it's time to put them outside. You can even get a leg up in planting *bare-root trees* (dormant trees whose roots aren't encased in soil — this is the common way to ship trees).

Setting up a glass (Or plastic) house

Traditionally, the greenhouse is a separate building, with walls of glass or plastic that let the sun come in and nourish the growing plants, but you can choose from many designs. You can even build a lean-to on the south side of your house or use an entire floor or room in the house as long as it gets good southern exposure.

Gather the materials yourself or go for a kit with everything already cut and sized and ready to put together. Here are some basic components of a greenhouse:

- ✔ **Frame:** Materials for the frames vary widely. You can go with a wood frame, a hoop design using PVC or metal conduit, fiberglass, or molded plastic.

- ✔ **Walls:** The walls can be glass, plastic, or polycarbonate panels (see-through rigid plastic). Glass is expensive, so plastic is a good alternative. A friend goes with a double layer of plastic film, which adds an insulation buffer. Glass is a poor insulator, so unless you want to heat the greenhouse during a cold winter, you typically use a glass greenhouse only in warmer months.

- ✔ **Heat source:** You can go with the simple form — just letting the sun do its thing during the day — or you can install solar-powered heating (see Chapter 7). Another heating option is to put a large compost heap inside. Compost generates heat as it decomposes, which can heat the greenhouse.

- ✔ **Water system:** Whether you simply use a hose or set up an automatic sprinkling system, you need some way to get water to the plants.

- ✔ **Ventilation:** You have to keep the temperature inside within the optimal growing range, not only in cooler seasons. Temperatures in a hot summer area can go over 120 degrees inside. To be sure the heat doesn't kill the plants, you need to ventilate. The air should contain enough moisture to grow plants but not so much that rot and mildew form.

Basically, use what you have available and get into it based on how much you want to spend. You can spend thousands on a state-of-the-art greenhouse, but do you really need to? And maybe you can have a year-round greenhouse, but if it costs $300 a month to heat it, do you really spend that much on a month's worth of vegetables?

Enjoying that southern exposure: Using your greenhouse

Greenhouses are great for starting seeds early so the plants are ready to go in the ground at the very start of the growing season. When you start seeds varies, but a good rule is to plant seeds six to eight weeks before the average last killing frost for your area. You may adjust that number if you have a very short or long growing season. Here's how to get your seedlings ready:

1. **Plant the seeds.**

    Take a large undivided flat, fill it with a couple of inches of soil mix, and tamp it down. Then sprinkle an entire packet of seeds over the mix, tamp them down into the mix, and water. Let them sprout. I do this with tomatoes and peppers, but other vegetables (such as cabbage) go into individual pots.

    For air circulation, run fans in the greenhouse. A slight breeze causes the plants to wiggle and encourages tougher stems.

2. **When you see the first true leaves, pull the seedlings out of the flat and put the strong ones into individual little pots, being careful not to break roots.**

    Add some potting soil and give the plants lots of light. They'll also need to be kept moist (but not wet, or they'll develop root rot).

3. **Harden off your plants.**

    About a week before planting, start putting the seedlings outside for several hours a day, and bring them inside during cool nights. Gradually increase the time the plants spend outside.

4. **Put the plants into the ground after the last predicted killing frost.**

    Find predicted frost dates from seed companies or your weather service.

You may also consider using the greenhouse to grow plants that don't grow well in your region, such as tropical plants like bananas, pineapples, oranges, lemons, or key limes. I experimented with pineapples. (It actually took five years to get the right combination of soils, pot size, and care. I ended up with a tiny, four-spear fruit that was all gone after three bites. But man, was it delicious! And the wonderful pineapple fragrance hit me in the face every time I entered the greenhouse.)

Part V

Using the Fruits of Your Labor

"..and here's what we canned from the garden this year. Beets, carrots, cucumbers...oh, there's that glove I couldn't find."

In this part . . .

Now that you have these beautiful fruits, vegetables, and animal products, you probably want to put them to good use — whether as sustenance for your own well-being or as a means to help maintain the farm. Imagine making your own cheese, yogurt, or jam. Discover how to best deal with meat products and even what to do with animal poo. Find out how to process wool and fiber and then use it in a variety of crafty projects. These chapters give you some creative ideas about how to use the products you get from your farm.

Chapter 15

Preserving and Using Food Items

. . . .

In This Chapter

▶ Storing grown foods for the short term

▶ Preparing fruits and vegetables for long-term preservation

▶ Making some money from your surplus

▶ Digging into the recipe box

. . . .

**Recipes in
This Chapter**

▶ Queso Fresco (Fresh
Cheese)

↻ Homemade Yogurt

↻ Cinnamon Applesauce

↻ Fresh Salsa (Salsa
Fresca)

▶ Todd's Spaghetti Sauce

↻ Todd's Zucchini Bread

🦃🥚↻🌶🥕

F resh fruits and veggies are wonderful, but the
bad news is that they pretty much grow only
during warm weather, and they don't last forever. If you want to continue to
eat the stuff from your own garden year-round, you have to do something
to preserve them.

In this chapter, I discuss produce storage and preservation, including info on
root cellar setup and maintenance. You find information on selling your plant
products at roadside stands or farmer's markets. Also in this chapter, I share
some time-tested, oldie-but-goodie recipes for dairy products and produce to
help you enjoy your foods in even more ways.

The Cellar: Getting to the Roots of Short-Term Storage

Before refrigerators came into the world, people used another method of
keeping foods cool: an underground room known as a *root cellar*. Even in
today's world of modern conveniences (or especially in today's world of
disappearing resources), a root cellar can help extend food's shelf life without using electricity or releasing *chlorofluorocarbons* (CFCs), ozone-depleting
gases that were used in refrigerators before being internationally banned
in 1996.

Besides keeping foods cool, root cellars also keep stored foods from freezing in the wintertime. In this section, I discuss how to put storm cellars to good use. (For info on construction and design, flip to Chapter 5.)

Using a root cellar: The underground basics

The stability of the temperature depends on how deep the cellar is. If you can swing it, you can maintain a constant temperature ideal for storage at 10 feet. Here are some ways to get the best use out of the cellar:

- ✔ **Maintain humidity.** Humidity is especially important if you live in a dry climate. Giving the cellar a dirt floor helps maintain moisture in the air. Another tip is to put in an open source of water, such as a small tub or pan, on that dirt floor.

- ✔ **Cover veggies with a cotton cloth or some other absorbent material.** If the humidity gets too high, condensation can form, introducing moisture that can harm the stored veggies. To help protect vegetables from this moisture, cover them with something like cotton cloth or some other absorbent material.

- ✔ **Make sure the room is ventilated.** Good air circulation is important in keeping the atmosphere where it should be.

- ✔ **Keep the door closed as much as possible.** Limit your trips into the room by planning ahead and taking out more than one thing at a time. Each time you open the door, you're letting a little of the outside environment in, which can mess with the temperature and humidity.

- ✔ **Rig up some electricity and keep it warm in the winter.** During very cold days, keep a light bulb on to help ward off the too-cold air.

Storing specific foods in the root cellar

Vegetables need a little preparation before you put 'em in the cellar. Cut leaves and stems off, but don't cut into the flesh of the vegetable, which promotes spoiling. Refer to Table 15-1 for some preparations specific to the vegetable.

Pick veggies at optimum maturity, not before they ripen or after their prime. Ripe vegetables have the best flavor and the highest nutritional content. Choose only those veggies in perfect shape, and remove any items from storage if and when they do start to spoil, because this spoiling is catching. (Damaged or ripening produce gives off *ethylene,* a gas that encourages other fruits and veggies to further ripen, so yes, one bad apple really can spoil the whole bunch.)

Table 15-1		Storing Various Veggies	
Vegetable	Special Preparations	Optimal Temperature/ Humidity	Storage Life
Brussels sprouts		32–40 degrees 90–95% humidity	3–5 weeks
Cabbage		32–40 degrees 80–90% humidity	3–4 months
Carrots	Put them in cold storage as soon as they're harvested.	32–40 degrees 90–95 % humidity	4–6 months
Onions	After clipping the tops off, keep them in the sun for a week before storage.	35–40 degrees 60–70% humidity	4–6 months
Peppers (sweet)		45–55 degrees 85–90% humidity	2–8 weeks
Peppers (hot)		50–60 degrees 60–70% humidity	2–8 weeks
Potatoes	Toughen the skins before storage by letting them sit in the open air (but not in the sun) for 2 weeks.	32–40 degrees 80–90% humidity	4–6 months
Radishes	Put them in cold storage as soon as they're harvested.	32–40 degrees 90–95 % humidity	2–3 months
Squash	Leave the stem on and let the squash sit in the sun for about 2 weeks, until the rind hardens.	50–60 degrees 60–70% humidity	4–6 months
Tomato (ripe)		45–55 degrees 85–90% humidity	
Tomato (green)		50–70 degrees 60–70% humidity	4–6 weeks

Preserving Foods for Long-Term Storage

You can't always go out back and pick a fresh tomato or apple, so if you want to be able to enjoy the fruits of your labor all through the winter, you need to do some preserving. *Canning* (sealing the foods into glass jars) is a great way to extend the shelf life of fruits or veggies. Some vegetables can be fermented. And don't forget about those delicious jams and jellies that are staples at breakfast tables across the world. This section covers all these methods as well as pickling, drying, and freezing.

Harvest and preserve foods at their peaks. Starting with the fruit or vegetable at its best goes a long way to giving you a tasty product.

Canning — the jarring effect

Canning uses a special jar that has a two-piece metal lid — a screw-on metal band and a flat lid with a rubber seal. This makes the jar airtight, which prevents spoilage caused by bacteria and mold.

A large stock pot may work for canning high-acid foods (such as fruits, salsa, and pickles), but for low-acid foods (such as vegetables), you have to use a pressure canner (see Figure 15-1).

Figure 15-1:
Use a pressure canner for preserving vegetables and stews.

Here's the basic canning process:

1. **Prepare the food according to the recipe.**

   Some foods need to have an acid such as lemon juice or vinegar added to the mix to aid in preservation. Some foods need to be peeled, and some need salt added. Most need to be blanched (boiled for a short time and dipped in ice water) prior to canning.

2. **Warm the jars and lids in hot (not boiling) water.**

3. **Remove the hot jars from the water and fill them with your food; place the flat lids on top and screw on the metal bands.**

4. **Boil the jars in a large covered stock pot or canner for the right amount of processing time.**

   Processing time varies according to food and altitude. The jars should be completely covered in water.

5. **Turn off the heat, remove the pot's lid, and let the jars sit for 5 minutes.**

6. **Remove the jars from the water and let the jars sit for 12 to 24 hours before checking the seals.**

The Natural Center for Home Food Preservation (www.uga.edu/nchfp/how/can_home.html) can give you details on the canning procedures for a large list of fruits, vegetables, nuts, and even meats.

Making jams, jellies, preserves, and marmalades

Depending on the way the fruit is processed, you can end up with jam, jelly, preserves, or marmalade. These luscious spreads are made from fruits that have been thickened, or jellied. Here's how these spreads differ:

- ✔ **Jams:** Jams are made from crushed fruit.
- ✔ **Jellies:** Jellies come from fruit juice.
- ✔ **Preserves:** Preserves involve chunks of fruit added to a jam base.
- ✔ **Marmalades:** Usually made from citrus fruits, marmalades include pieces of the rind with chunks of fruit in a jam base.

But these treats aren't just for fruit! Try making jelly out of your peppers, mint, basil, or other herbs to serve with cream cheese and crackers or use as a meat glaze!

Jam should be made in small batches (around 6 cups at a time), or the solution won't set. Here are the basics for making a fruit jam:

1. **Wash and peel the fruit, cut it into small pieces, and crush those pieces.**

   For marmalade, use a citrus fruit and leave some of the rind in. For preserves, don't crush the fruit.

2. **To prevent the fruit from turning brown, add a little lemon juice.**

   For jelly, instead of adding lemon juice, put the fruit into a large pot, add about 1 inch of water, and boil until the pieces are soft. Then use a jelly strainer or cheesecloth to drain the liquid from the pieces. Use this liquid for the rest of the steps.

3. **Add sweetener according to the recipe.**

   For a lower-calorie version, try using Splenda or canned fruit juice.

4. **Add pectin, the stuff that makes the jam stiffen.**

5. **Heat the mixture to a boil, and boil it for 1 minute.**

6. **Put the jam into canning jars and can (see the preceding section).**

You can also find recipes for a no-cook freezer jam.

Fermenting and pickling

Fermentation converts sugars or carbohydrates to alcohols. Everyone knows about fermentation as related to turning fruit juices into wine or starches in grains into beer. But did you know that sauerkraut (and the Korean staple kimchi) is fermented cabbage? The fermented version of cabbage is even healthier than the raw version. Sauerkraut contains all sorts of good stuff. Studies suggest it boosts the immune system, helps prevent cancer, and helps your digestive system run smoothly.

Pickling involves soaking the veggie in a solution of vinegar or brine (salt water). In the brine solution, the veggie ferments. Some vegetables need preparation before pickling. For instance, asparagus should be blanched first, and beets should be precooked in their skins for 30 minutes. On the other hand, you can raw-pack green beans, carrots, onions, mushrooms, and zucchini.

Cider vinegar is a good choice for the pickling solution, but light-colored veggies such as onions or cauliflower do better in clear vinegar.

Cucumbers aren't the only veggie that gets pickled. You can pickle most vegetables and even eggs! Meat stakes its claim in the pickling world, too — ever tried pickled pigs' feet? However, many people like to call the pickling process *curing* when you use a brine solution with meat — see Chapter 16.

Drying

Drying fruits, vegetables, and meats reduces internal moisture, giving them a much longer shelf life, and it's one of the oldest methods of preserving foods. The old-fashioned way to do it was to simply let the food dry out in the sun. Today, you can get one of those fancy yet inexpensive dehydrators, or you can do it in the oven.

For everything to work correctly — that is, to dry things instead of rotting them — you need the following components:

✔ Enough heat to draw moisture out of the food without cooking it

✔ Dry air to absorb the released moisture

✔ Enough air circulation to whisk moisture away

Because a big component in foods is water, when you remove that water, you're left with a much smaller pile of food. That makes the food easier to store because it takes up so much less space. Table 15-2 shows how much dried food you get after drying, as related to the amount of starting produce.

Table 15-2	Starting Amounts and Quantity of Dried Produce	
Produce	*Fresh Produce (Pounds)*	*Dried Product*
Apples	12	1.25 pounds 3 pints
Beans, lima	7	1.25 pounds 2 pints
Beans, snap	6	0.5 pounds 2.5 pints
Beets	15	1.5 pounds 3–5 pints
Broccoli	12	1.375 pounds 3–5 pints
Carrots	15	1.25 pounds 2–4 pints
Celery	12	0.75 pounds 3.5–4 pints
Corn	18	2.5 pounds 4–4.5 pints

(continued)

Table 15-2 (continued)

Produce	Fresh Produce (Pounds)	Dried Product
Grapes	12	2 pounds 3 pints
Greens	3	0.25 pounds 5.5 pints
Onions	12	1.5 pounds 4.5 pints
Peach	12	1–1.5 pounds 2–3 pints
Pear	14	1.5 pounds 3 pints
Peas	8	0.75 pounds 1 pint
Pumpkin	11	0.75 pounds 3.5 pints
Squash	10	0.75 pounds 5 pints
Tomatoes	14	0.5 pounds 2.5–3 pints

Adapted from "Drying Fruits" and "Drying Vegetables," Preparation (Colorado State University Extension, 2003 and 2004), available at www.ext.colostate.edu/pubs/foodnut/09309.pdf and www.ext.colostate.edu/pubs/foodnut/09308.pdf.

Freezing

Some fruits and vegetables don't freeze well at all. The scientific explanation is that water expands as it freezes, breaking the fruit or veggie's cell walls and giving you a pile of brown mush as it thaws. Those fruits and veggies that make poor candidates for freezing include cabbage, celery, lettuce, tomatoes, and cucumbers — the typical salad veggies — as well as bananas and potatoes. (However, frozen paste tomatoes do work fine for stews.) Good freezing candidates include the following (anything you may see in the frozen food aisle in the grocery store):

✔ Brussels sprouts

✔ Peppers

✔ Corn

✔ Peas

- Beans
- Cherries
- Strawberries
- Blueberries

Before you freeze, some fruits and vegetables benefit from a *blanching* (steaming or boiling for about 1 minute). This application of heat destroys the enzymes that cause the produce to lose nutritional value, color, texture, and flavor. Blanching also serves to soften the vegetables and make them easier to pack. After the short heat bath, dip them in cold water to quickly cool them and stop the cooking process.

Squashes and sweet potatoes should be fully cooked before freezing. Onions, peppers, and herbs don't need to be blanched, but all other fruits and veggies do.

Selling What You Don't Keep

Even though this farming thing is a hobby for you, you can still make a little pocket change from the fruits of your labor. You can sell your wares at the local farmer's markets, at roadside stands, or directly to the top restaurants in town. Or you can let the customers come to you and do their own picking. Even if making money isn't a concern for you, these options offer a fun way to pass along any excess food you've grown.

Check with your local and state laws to make sure you stay in line with regulations concerning food sales to the general public. If you operate as a produce stand (you handle raw, unprocessed fruits and vegetables as opposed to selling prepared food), you may get away with simply having a food handler's certificate or refrigeration at your booth. If you're selling goods cooked or baked in your kitchen, you may have to have your kitchen inspected. Some states don't let you prepare and sell items considered potentially dangerous, including meat, fish, poultry, or dairy products, without a license.

Participating in farmer's markets

Setting up shop at a farmer's market is an easy way to get a little pocket change from the fruits and veggies on the farm. It's an income opportunity that doesn't require a huge cash outlay before the fact. Participating in a local farmer's market is also a good way to let those excess veggies go to a good home instead of rotting on the vine. You can sell the fruits and vegetables raw or in a preserved state. Your canned beans, strawberry jam, or homemade sauerkraut may be a big hit.

Depending on the rules at each particular market, you can rent space for a booth for one afternoon or for the whole summer. Your booth can usually be as plain or fancy as you want it to be. Although not usually required, a shelter is a good idea because you'll be standing in the elements for a few hours. Then you just need a table and/or boxes to display your wares and some plastic bags to make it easy for your customers to load up and carry their purchases home.

Selling to local stores and restaurants

If you grow produce beyond what you can consume, you may want to try selling it to local stores or restaurants. With the buy-local movement sweeping the country, a lot of food retailers are interested in finding local sources. Buying food close to its source means the foods are fresher because they don't have to travel too far to get to the store or restaurant. Buying local also means less impact on the environment (less gas for transportation, less in the way of packaging materials, and so on).

Each buyer is looking for different things. Some may want all-organic, others may just want good quality, others may want special veggie varieties, and others may be looking to get quality ingredients less expensively.

Ask around. Find out the name of the buyer at the local grocery store. Smaller mom-and-pop stores are easier to approach, but that doesn't mean your branch of a national chain won't be interested.

If you have a favorite restaurant, see whether the chef there is interested in your homegrown, pesticide-free fruits and vegetables. Ask to speak to the buyer or owner. If it's a place you frequent, maybe even chat with the wait staff and get a feel for whether they may be interested and whom you should talk to. Offer to hand out samples so the owner can see just how awesome your foods are.

After you've established a business relationship, treat the deal as any other business deal. Be consistent, provide the best products and service you can, and be sure you both understand the payment arrangement.

Manning roadside stands

Just like the old-fashioned lemonade stand, roadside stands with boxes of brightly colored fruits and vegetables can attract the attention of passersby. Set up late in the afternoon and catch people as they're heading home for supper. If you don't know what to charge, check with other roadside stands, wander around a farmer's market, or even just see what the local grocery store is charging.

Setting up one of these stands can also be a great way to get your kids involved in farm life — let them get their hands dirty by caring for and picking the things to sell, and then give them a cut of the profits.

You can spend some money on advertising in the paper or maybe a local radio show, or you can simply set up your stand and some signs to entice buyers. If you don't live near a main road, you can see whether a friend will let you set up shop on his or her sidewalk.

Consider the logistics of the actual operation. Here are some options:

✔ Sell items in a pre-designated quantity (a small or large box of apples) or individually by the pound or dozen.

✔ You want to keep yourself and the product sheltered, so either set up an umbrella or canopy over some tables outside or set up the tables in the barn.

✔ Maybe you want to trust a little and not man the booth the whole time it's open. Instead, have a bell or just watch for traffic. A farmer in my hometown sells eggs on the honor system. He usually doesn't even come out. You just take your eggs and leave a dollar in the slot.

 Be ready to answer questions about the product — customers may want to know what kind of pepper that is, whether it's hot, whether it's organic, and so on. You may want to cut a veggie into small pieces and offer samples to customers before they buy.

Setting up a U-pick operation

Maybe you don't have time or interest in sitting in a booth for several hours just to get a little extra pocket change. Why not set your own rules and hours, letting the customers come to you to harvest the fruits and veggies themselves?

You can charge by the basket, by the number of items, or by the pound. The best way to get an idea of what you should charge is to check at competing U-pick places. Checking roadside stands and farmer's markets can give you a ballpark amount, but those vegetables come with the value-added characteristic of the farmer's time and sweat doing the picking. Customers should get a price break if doing the picking work themselves. However, if you're offering something unique — the tastiest tomatoes in the county due to your hard work at carefully selecting what to plant and the most tender care while they grow — you may be able to charge a premium.

You have to advertise to let people know you have fruits or veggies for the picking. Put up a sign at the local country store or even at the grocery (many stores have an ad board by the front doors). Put a sign in your front yard or at a nearby busy intersection. Consider a phone book or newspaper ad, or even set up your own Web site. Create an ad on an online local classifieds board — at the very least, go to www.pickyourown.org and register online for free.

Creating New Treats from Homegrown Goodies

The list of what you can do with the plant and animal products is endless. Seriously! More than a couple of books could be — and have been — written to help you turn fresh farm products into things that are deliciously edible. This section offers you a few ideas and time-tested recipes.

These recipes give you options for creating dishes that can be eaten right away, or in some cases, stored by canning or other means for enjoying later. In the case of milk, the recipes offer a way to take a product that can spoil quickly and extend its useful life by turning it into cheese or yogurt. These recipes also yield products that you can sell in your roadside stand or at the farmer's market or give as gifts to friends and relatives.

Milking the dairy recipes

Generally, when people think of milk, they think only of the milk that comes from cows. But other mammals produce milk that's good not only for the animal babies who thrive and grow to maturity on it but also for humans. In fact, outside of the U.S., goat's or sheep's milk is more popular than cow's milk. Any mammal who can be milked can be a potential source of the nutritious stuff.

Besides drinking milk, you can make cheese, yogurt, butter, and ice cream, to name just a few ideas. In fact, recipes for cheese and yogurt are coming right up! You can make both without expensive equipment, though you can buy a gadget such as a cheese press if you really get into it (see Figure 15-2). Although people have developed hundreds of types of cheese, they all have the same basic ingredients and go through the same process.

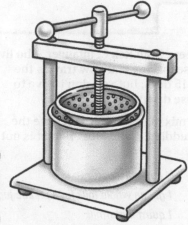

Figure 15-2:
A cheese
press.

Some applications, such as yogurt, require that your fresh-from-the-cow (as opposed to store-bought) milk be pasteurized before use. Pasteurization (heating and then cooling) is a process that aims to kill off most of the dangerous disease-causing bacteria in milk without significantly changing the milk's flavor or nutrition content.

Commercially prepared milk is typically pasteurized at a very high temperature (285 degrees) for a very short time (2 seconds). You can do a home version of the process, but heating the milk does change its flavor somewhat:

1. **Clean and sterilize empty milk bottles and tops by boiling them in water for 10 minutes.**

2. **Pour the raw milk in the top part of a double boiler and fill the bottom pan about half full with water.**

3. **Heat the milk slowly until it reaches 145 degrees and keep it at this temperature for 30 minutes.**

   Periodically place a cooking thermometer in the milk — don't leave the thermometer in where it comes in contact with the side or bottom of the double boiler. Stir often to prevent burning and to keep the temperature even. You'll likely have to adjust the heat setting to keep the temperature from fluctuating away from 145 degrees.

4. **Cool the milk quickly by putting the top part of the double boiler in ice water.**

   Stir often to make it cool off faster.

5. **Pour the cooled milk into the sterilized bottles and put them in the fridge.**

Queso Fresco (Fresh Cheese)

This is a basic recipe for a squeaky white cheese from Rex C. Infanger, who lives in Utah. He traveled the world studying cheese-making and now travels the world consulting. This cheese lasts for up to 15 days, but you don't have to worry about that because this delicious cheese disappears quickly!

You can find Junket brand *rennet* tablets — a mix of enzymes that cause the milk to curdle — in grocery stores near the pudding. *Note:* Most rennet is not vegetarian.

Tools: *Bleach, two 6-quart saucepans, colander, thermometer, cheesecloth*

Preparation time: *2 hours*

Yield: About 1 pound (16 servings)
1 rennet tablet

½ cup cold water

1 gallon whole milk (in a plastic jug)

1 quart buttermilk

1 teaspoon salt (not iodized or sea salt)

1 Sterilize your equipment with a mild bleach and water mixture. Prepare the draining pot by placing a 6-quart saucepan in the sink. Either put a colander in the pot and line that with cheesecloth or place cheesecloth over the top of the pot (make sure it has a little slack in it). You may need to secure the cheesecloth over the pot with an extra large rubber band.

2 Dissolve the rennet tablet in the cold water for about 10 minutes (follow the manufacturer's directions). Make sure it's totally dissolved, giving the water a stir to get up any undissolved chunks.

3 Mix the milk and buttermilk in a 6-quart saucepan and heat the milk mixture to 90 degrees, using a thermometer to get the temperature right. Remove the pan from the heat and add the dissolved rennet to the milk mixture; stir for 1 minute.

4 Allow the rennet/milk mixture to sit with no stirring until a firm coagulum (clot or curd) forms, usually about 25 minutes. The curd is ready to cut when it shows a clean break. Test for this by inserting your thermometer at an angle and then lifting it; the curd should separate and leave an open mark on the surface. *Note:* If the curd takes longer than 45 minutes to show a clean break, add slightly more rennet the next time you make your recipe.

5 Using a long knife, cut through the coagulum at an angle (going lengthwise one time, widthwise the next) to get approximately ½-inch cubes. Over medium heat, cook the coagulum-liquid mix to 110 degrees while stirring (this should take about 7 minutes but no more than 15 minutes). Remove the mixture from the heat and keep stirring a little bit as it continues to thicken.

6 Pour the mixture onto the cheesecloth in your draining pot so the *whey* (the liquid part) soaks through into the other saucepan and the cheese curds stay in the cheesecloth.

7 Wrap the cheesecloth around the curds. You can squeeze it a little, too. Put the cheesecloth-wrapped curds in the colander and place the colander in the empty saucepan. Fill the gallon jug your milk came in with 105-degree water and place it on top of the wrapped curds for 10 minutes to let the remaining whey come out (and fall through the colander and into the pan). Remove the jug, unwrap the cheese, and cut it into ¾" x 2" strips.

8 Sprinkle the salt on the curd you've cut up and gently mix. Then put the cheese back in the cheesecloth and wrap it tight enough to form a ball. Refill the jug with 105-degree water, place it on top of the cheese ball in the colander, and allow the cheese to set. Press the cheese for 30 minutes. It's done! If you don't eat it all immediately, store it in the fridge.

Vary It! *You can also toss in some herbs when you sprinkle the salt on the curd. Mr. Infanger doesn't like to use herbs in his recipe, but I can't resist tweaking recipes. I made a batch with some basil, and it was yummy.*

Tip: *The fresher the milk, the better the quality of the cheese. I use whole goat's milk (whole milk has the best flavor and texture, whereas skim milk gives a tough texture with little flavor).*

Tip: *The leftover whey has a good flavor and is full of nutrients. You can use it in other recipes, such as in breads. I mix it with ice and pineapple in a blender. Although a touch cheesy, it's delicious, and if you didn't know where it came from, you'd never question the source of the tasty drink.*

Per Serving: Calories: 181; Total fat: 9.1 g; Saturated fat: 5.3 g; Cholesterol: 29 mg; Sodium: 296 mg; Carbohydrates: 14.3 g; Fiber: 0.1 g; Sugar: 16.1 g; Protein: 10.4 g

🍅 Homemade Yogurt

Yogurt is the result of fermentation. It contains some ultra-beneficial bacteria, and it's very good for you. You can make on your own yogurt using either store-bought milk or pasteurized milk from your own cow or goat. Commercially bought yogurt works as a starter so you can get your own cultures going — make sure the label says the yogurt contains a live culture.

Preparation time: *1 hour*

Incubation time: *8 to 14 hours*

Yield: 4.5 cups (9 servings)1 quart whole milk

⅓ cup dry milk powder

¼ cup commercial plain yogurt

honey, sugar, fruit pieces (optional)

1 Pour the milk into a pot and heat on low, stirring occasionally with a metal spoon. Check the temperature with a thermometer. When the temperature reaches 180 to 185 degrees, take the milk off the burner.

2 Put the plain yogurt in a small bowl and set it on the counter. Let the milk cool at room temperature, stirring and checking the temperature frequently. When the temperature gets back down to 105 to 110 degrees (which should take 10 to 20 minutes), stir in the dry milk powder and the plain yogurt.

3 Put the mixture into smaller containers and cover them tightly with lids. Let the yogurt incubate for 8 to 14 hours, keeping the temperature between 105 and 122 degrees. Maintain this temperature by putting the containers in your oven and turning on the pilot light or placing the containers in pans of warm water. The longer the mixture incubates, the tangier the yogurt.

4 When the yogurt has the right consistency and tanginess, put it in the refrigerator and let it cool for several hours before serving it. If desired, add honey, sugar, fruit pieces, jam, or other flavorings.

Vary It! *Any milk will do — cream, whole, skim, goat's milk, or even soy milk. The higher the fat content, the creamier and smoother the yogurt. If you use skim milk, increase the amount of dry milk powder to ⅔ cup.*

Per Serving: Calories: 93; Total fat: 5 g; Saturated fat: 2.9 g; Cholesterol: 16 mg; Sodium: 64 mg; Carbohydrates: 7 g; Fiber: 0 g; Sugar: 7.8 g; Protein: 4.9 g

Making forays into fruit territory

Although most people who live on farms come up with delicious recipes by experimenting with their fresh fruits, seeing what other people have come up with is always fun. Here's a recipe that I come back to time and again.

⬤ Cinnamon Applesauce

I like the Granny Smith or McIntosh varieties for applesauce, but your preferences may vary. Serve your applesauce cold or warm.

Preparation time: 30 minutes

Cooking time: 30 minutes

Yield: 1 quart (8 servings)
3 pounds apples

(about 5 medium apples)
⅔ cup sugar
1 teaspoon ground cinnamon
1 cup water

1 Peel, core, and slice the apples into thin slices.

2 Place the apple slices in a large pot. Sprinkle the sugar and cinnamon over the apples and then add the water. Bring the mixture to a boil.

3 Reduce the heat. Cover and simmer the apples for 20 minutes.

4 Using a potato masher or a food processor, mash the apple pieces to the desired consistency.

Tip: Adding a dash of lemon juice helps retard the browning of the apples during the cooking process so the end result is a light tan color.

Per Serving: Calories: 113; Total fat: 0.1 g; Saturated fat: 0 g; Cholesterol: 0 mg; Sodium: 0 mg; Carbohydrates: 29.6 g; Fiber: 1.5 g; Sugar: 26.7 g; Protein: 0.3 g

Eating all your veggies

Although most people who grow vegetables do so because they love to eat veggies, not everybody shares the same appreciation. Instead of putting raw or steamed veggies on the table, entice your family with some of these delicious foods.

🍎 Fresh Salsa (Salsa Fresca)

Fresh salsa made from fresh tomatoes is out of this world. Serve fresh salsa as a dip with chips, on tacos or fajitas, or as a garnish on meats.

Preparation time: *15 minutes*

Yield: *3.5 cups (7 servings)* *2 pounds tomatoes (about 3 medium tomatoes)*

1 small yellow onion

½ cup chopped fresh cilantro

¼ teaspoon minced garlic

1 serrano chile or jalapeño pepper

¼ teaspoon salt

Dash of lime juice

1 Dice the tomatoes and onion, and mix them with the chopped cilantro and garlic in a bowl.

2 Remove the seeds and stem from the hot pepper, and chop the pepper finely. Add the pepper until you get the desired flavor and heat.

3 Add the salt and a dash of freshly squeezed lime juice to taste.

Vary It! *If you prefer a milder salsa, substitute anaheim chilies or green bell peppers for the hot peppers.*

Per Serving: Calories: 17; Total fat: 0.1 g; Saturated fat: 0 g; Cholesterol: 0 mg; Sodium: 87 mg; Carbohydrates: 3.8 g; Fiber: 1 g; Sugar: 2.1 g; Protein: 1.1 g

Todd's Spaghetti Sauce

Spaghetti sauce is an old standby. You can use it on spaghetti, lasagne, or pizza or even as an accompaniment to baked potatoes. This recipe came about from years of trial and error. We use Roma tomatoes, but you may want to experiment with some of the varieties you grow.

Preparation time: *15 minutes*

Cooking time: *30 to 40 minutes*

Yield: *About 3 cups (6 servings)*
1 pound pork or ground beef, or 4 chicken breasts

2 tablespoons olive oil

2 tablespoons chopped, fresh basil (or ½ teaspoon dried)

1 clove garlic, minced

1 small yellow onion, chopped

½ green bell pepper, chopped

4 large mushrooms, chopped

4 ounces (½ can) black olives, chopped

4 large tomatoes, peeled, seeded, and puréed in a blender

12 ounces tomato paste (two 6-ounce cans)

Salt and pepper to taste

1 Fully cook the pork, chicken, or ground beef, using whatever method you prefer; drain off the fat if needed and set it aside. Pan-frying can give the meat a browned skin, and baking is the healthiest but longest process. For chicken, you can use the microwave if desired.

2 In a large skillet, heat the olive oil, basil, and garlic over medium heat. After 1 or 2 minutes, add and sauté the onion, bell pepper, mushrooms, and olives. Cook the veggies for about 2 minutes, until they're slightly softened.

3 Add the puréed tomatoes and tomato paste, stirring to make the mixture uniform. Then add the cooked meat. Simmer the mixture for 15 to 30 minutes. Add salt and pepper to taste.

Tip: *Put the sauce in the refrigerator and serve it the next day. For some reason, all these different components actually improve with age.*

Per Serving: Calories: 129; Total fat: 6.4 g; Saturated fat: 0.7 g; Cholesterol: 0 mg; Sodium: 537 mg; Carbohydrates: 18.4 g; Fiber: 4.3 g; Sugar: 11.1 g; Protein: 4.2 g

☙ Todd's Zucchini Bread

You wouldn't think a squash would be so good as a dessert, but it's true!
Zucchini adds moistness, just like the carrots in carrot cake. Zucchini bread is a
great way to use up all those extra zucchinis.

Preparation time: *15 minutes*

Cooking time: *50 to 60 minutes*

Yield: 2 loaves (20 servings)

3 eggs, beaten

1 cup vegetable oil

2 cups sugar

1 teaspoon salt

½ cup walnuts, chopped

2 cups zucchini, peeled and grated

3 teaspoons vanilla

3 cups plus 2 tablespoons flour

1 tablespoon cinnamon

½ teaspoon baking powder

1 teaspoon baking soda

1 Preheat the oven to 325 degrees. In a large bowl, mix the eggs, vegetable oil, sugar, zucchini, and vanilla.

2 In a medium bowl, stir together the flour, cinnamon, baking powder, baking soda, and salt; add these dry ingredients to the egg mixture, mixing until uniform (this should take about 2 minutes on low speed with a hand mixer). Stir in the nuts.

3 Pour the mixture into two oiled 8.5" x 4.5" x 2" loaf pans. Bake the bread for 50 to 60 minutes or until a toothpick inserted in the center comes out clean.

Per Serving: Calories: 282; Total fat: 14 g; Saturated fat: 1.8 g; Cholesterol: 32 mg; Sodium: 203 mg;
Carbohydrates: 36.1 g; Fiber: 1.1 g; Sugar: 20.4 g; Protein: 3.5 g

Chapter 16

Where's the Beef? From Hoof to Table

● ●

In This Chapter

▶ Finding a reputable meat cutter

▶ Selling the whole animal

▶ Preserving meat

● ●

Home-raised meat doesn't compare to what you find in stores — it's so much better! However, raising your own meat is trickier than growing your own fruit or veggies. Pulling a tomato off the vine and popping it into your mouth isn't so hard, but getting food that comes from your critters into table-ready condition takes a lot more work.

Used to be, farmers just prepared animals themselves, but these days, it's usually done by a professional. You have to follow restrictions and standards — especially if you're selling the meat to the general public — to ensure the meat is safe and healthy. As a small farmer, you may want to avoid going through those hoops and instead use a reputable meat cutter. Or you can forego being part of the butchering process entirely and simply sell live animals. In this chapter, I discuss finding a meat cutter, selling animals, and preserving meat.

Be aware that what I mean by *preparing* an animal is transforming an animal from a walking, four- or two-legged beast into one that becomes dinner. It's not for the faint of heart. I personally can't dress out an animal, especially one I raised. I recommend taking an animal to a professional.

Hiring a Reputable Meat Cutter

Just as in any other business, you have ways to find out who in the meat preparation business is doing a good, honest job and who's just trying to make a buck without putting in the effort. In this section, I give you information finding the best meat cutter for your business. I also provide info on doing a trial run and choosing the right animals.

Only reputable butchers are allowed to prepare meat that'll go to people other than themselves. These butchers have gone through the hoops with the USDA: They're trained and licensed by the state, they have a commercial kitchen, the state has inspected their conditions and certified them as sanitary, and so on. If the meat is going to be for yourself, you're allowed by law to dress it out yourself for home consumption. You can then cook and serve the meat to friends, but you can't sell it.

There's a difference between a slaughterhouse and a meat cutter (formerly known as a *butcher*). A *slaughterhouse* is a place where animals are killed, usually a lot of them at once and in an assembly line sort of fashion. The carcasses are sealed and boxed and sent to the meat cutter. The *meat cutter* (butcher) is the one who carves the meat into the various cuts. However, in rural areas, you may find an operation that does both the slaughtering and cutting in the same place on a smaller scale.

Letting your fingers (and feet) do the walking

In so many cases, the best relationships come by word of mouth. Ask your neighbors which meat cutters they've dealt with and what the experience was like. Walk out to the back fence and say, "Hey, we have a cow we want to take in. Know of a good place?" When I first met one of my neighbors and went into a beef business with him (he grew the beef and I found buyers for the meat), he had some definite opinions about a couple of the local meat cutters.

One good place to start finding a meat cutter (besides word of mouth) is the Yellow Pages, whether in print or online. Typically, if you look up "butcher," or "meat processing," you find that most meat cutters do their own slaughtering as well, though sometimes they work with another business that does the slaughtering. Here are some other options for finding meat cutters:

✔ **Take advantage of random findings.** Maybe you notice the local grocery store has some pretty amazing cuts (and otherwise qualities) of meat. Why not ask the meat cutter whether he or she does independent work or, if not, for a recommendation of someone who does?

✔ **Check out the local business-seekers.** Most hunters don't have the skills to process animals themselves, so they find somebody to do it for them. Anyone the hunters use to process wild game should be able to process any carcass, despite its domestic classification.

✔ **Peruse bulletin boards.** Bulletin boards in the local farm supply stores are good places to find all sorts of businesses related to farm operations. Check to see whether anybody's advertising butchering. Maybe even ask the employees at the farm store whether they know of anybody.

Meeting your meat cutter face to face

The best thing to do, after all the research or word-of-mouth references, is to simply walk into the business that interests you most and have a good old-fashioned face-to-face talk with the employees or proprietor. You want to make sure the meat cutter's work is going to mesh with what you want done.

Visit the shop and see what it looks like. Is it clean? Ask the proprietor your questions. You want to be sure he or she can do custom cuts to your specifications. For instance, maybe you want steaks cut 1 inch thick and two to a package, or perhaps you want roasts in 4-pound packs. If the meat cutter you initially talk to isn't set up to do custom cuts, ask whether he or she knows another meat cutter who can.

You also need to find a meat cutter who's able to work with the meat you have to offer. Some businesses work with the basic animals but not with anything considered exotic. *Exotic* is a subjective term, and some meat cutters want to do only the more popular animals, such as cattle, sheep, or pigs.

Not what he ordered: The beef swap

You expect that when you take something in to be processed, the butcher is going to give back what's yours. But unfortunately, that's not always the case.

One particular butcher turned out to be not-so-trustworthy. "Farmer Bob" took three almost identical steers in to be dressed out. There was probably a 50-pound difference between the heaviest and lightest. However, the meat he got back was quite a bit different. The output of one of the animals was a full 200 pounds lighter than what he turned over to the butcher, and the flavor and texture of the meat was so different from the other two that he knew something was amiss. He suspected that he got the meat from somebody else's old cow and that someone else got the meat from his prime steer.

You'd think a business that continually does this sort of sloppy organization wouldn't stay in business for long, but the person who got Farmer Bob's rightful meat is probably now a lifelong customer ("Wow, I really got some great meat from that butcher!"). Whether it's sloppy paperwork or something more sinister — such as taking a package of meat here and there because the customer won't miss it — make sure you steer clear of the slimy businessperson. The best way to avoid this kind of indignity is by word of mouth or by checking with the local Better Business Bureau.

Do the prices seem reasonable? The butcher my friend uses charges $40 to slaughter (including the fee for USDA inspection) and then 35 cents a pound to cut and wrap, but your costs will obviously vary.

Spend a few minutes talking to the meat cutter. Does he or she seem like somebody you'll work well with?

Doing a trial run

Even after you meet your meat cutter and visit the shop, you still have to do a trial run. The meat cutter may be reputable, and perhaps your conversation revealed that the meat cutter is probably going to be good to work with. But the final okay doesn't happen until you do your first business transaction. Were you satisfied? Did the meat cutter do what was promised? Did you get back the meat you brought in? Was the fat trimmed to your satisfaction? A not-so-good meat cutter may not trim enough off or may trim too much (certain cuts need a band of fat to enhance the flavor).

When you bring in an animal for processing, make sure you have proof of ownership. A member of the meat police — that is, a reputable, USDA-inspected dealer — pays attention to the animal's brand. If the animal has a legitimate brand, anyone wanting to process the meat has to have a bill of sale or proof that the brand is his or her own. (This practice goes way back to the Old West days, when cattle-wrangling was a common although law-defying way of life.) Nowadays, a brand inspector comes out and takes a look at the brand. Then, barring any discrepancies, the inspector lets the processing proceed if the animal isn't diseased.

As you decide which animal to send to the meat cutter, remember that only sound, healthy animals should make it to the dinner table. You don't want to eat an animal who was sick or diseased. Think of mad cow disease, leptospirosis (which causes flu-like systems), botulism, and salmonella. Even if the animals received antibiotics, you have to observe withdrawal times on food-producing animals before the meat is safe. If an animal contracts something nasty that can't be treated and fixed, have the critter put down and buried, burned, composted, or otherwise disposed of properly. Although this may sound harsh, remember that diseases can spread to other members of the herd as well as to humans or dogs.

Selling the Whole Animal

One way to get around the restrictions and procedures of having meat prepared for sale is to sell the whole animal intact and let the customer dress it out or find his or her own favorite meat cutter. The customer may ask you for a recommendation.

Not having the stomach for it

My first encounter with the butchering process left me (and my father) traumatized. We raised rabbits when I was a kid. I knew in the back of my mind just what went on, but I was never a part or it. I ate the meat and just distanced myself from where the meat on my plate came from.

One day, I went with my father to do errands. One of the stops was to drop off a few rabbits at the butcher. Still, I wasn't thinking about the reality of it and just assumed we'd drop off the animals and be on our way. The butcher came over to our car and picked up one of the rabbits, and as I was watching, he bashed it on the head and it went limp.

I was stunned. For a few seconds I couldn't move. Then I burst into tears. My father had to drive the car to another area, distancing me from the place where I saw the (in my mind) brutal death. I made my father cry, too.

I've never been able to watch a butchering, let alone do it myself, because I always remember that day when the reality of the process hit me like a gut punch. My farmer friend calls me a hypocrite because I have no trouble eating meat (chicken, beef, pork, and even an occasional rabbit stew). But in my mind, if I don't see it, I can handle it. Out of sight, out of mind.

My friend the do-it-yourself butcher often sells animals and helps the buyers with the meat preparation. He shows them what they need to do, and they then do it. He sometimes does demonstrations for potential buyers who want to do their own butchering. A lot of people say they enjoyed the experience and that preparing their own animals made them feel closer to dinner, having more of an appreciation for where the food comes from. It gave more of a feeling of pride of ownership.

However, witnessing a slaughter can also go the other way. He tells me of an 8-year-old boy who came to his farm with his parents to watch a turkey slaughter. At first the boy was excited to see how it was done, but by the time the job was finished, he was pretty traumatized, and his parents say that it was several years before the boy would eat turkey.

Often, a large animal is too much for one family, so people share the meat. I've shared meat many times. A whole cow is way more than my small family can eat, so I order an eighth. (That's really a half of a quarter — the butcher divides the meat down to a quarter, and then we do the rest of the splitting.)

Preserving and Flavoring the Meat

Before the days of refrigerators, people had to do something to keep the meat from spoiling, and these methods still work today. You can just put the meat in the freezer or dry it, or you can prolong the time before it spoils (as well as its quality and taste) by curing or smoking, which I discuss in this section.

Smoking

Smoking improves the flavor and makes the preserved meat less likely to spoil. The heat also helps dry the meat. The preserving ingredient in this method obviously comes from smoke, usually from burning hardwoods such as hickory, oak, or mesquite. This method takes about 24 hours in a smokehouse to accomplish, with three phases to the process:

1. **In the drying phase, heat the smokehouse to 125 degrees for 8 hours.**

2. **Introduce the smoke and increase the temperature to 135 degrees for another 8 hours.**

3. **In the final 8 hours, continue the smoke and increase the temperature to 180 degrees.**

Some of the fancier smoking treatments can get to be as expensive as the meat itself. Check out www.smoking-meat.com for all sorts of tips and recipes.

Curing

Curing meat is another option for preserving meat, and it's done by putting the meat in a solution of salt and other substances, such as nitrates and sugar. It can greatly improve a meat's flavor. I tried some uncured bacon once, and I can't even describe the flavor, but it was definitely not what I expected.

Although you can find many recipes, the two basic ways to cure meat are through soaking it in a brine bath and dry curing, or *corning*.

Salt-curing in brine is one of the oldfangled ways to preserve meat, having been done long before fancy gadgets came around. You can cure meat using only a container with a lid to hold the brine solution and the meat. Here's a basic method for curing meat:

1. **Fill your container with a solution that's 1 cup salt per 2 gallons of hot water and a dash of vinegar.**

2. **Cut the meat into large pieces (about 10-pound chunks) and add it to the brine solution.**

3. **Let the mix soak for 6 days.**

4. **Remove the meat, dry it off with cheesecloth, and place it in a gunny sack (burlap bag) to keep the flies off.**

5. **Hang the meat in a cool place to dry.**

It can last throughout the winter or up to 6 weeks in the summer.

Chapter 17

Preparing Fiber and Making Crafty Goods

In This Chapter
▶ Shearing animals
▶ Creating with animal fibers
▶ Making fiberless crafts

The fleeces from fiber animals such as sheep, goats, and alpacas can yield some beautiful and unique garments or other knit items. It's so rewarding not only to have a beautiful sweater to wear but also to have people ask where you got such a lovely thing, to which you can answer, "I made it myself. I even made the yarn, and this is from my llama Oggie!"

This chapter takes the fiber-animal owner through the process of getting the fiber off the critter and then getting the fiber into a form you can use in various applications, such as spinning, knitting, or weaving. I also discuss a few other fun crafts that use non-woolly farm products or byproducts. A lot of materials that seem like they're probably useless really do have some creative applications.

Shear Brilliance: Collecting and Processing Fiber

If you have fiber animals (alpacas, some breeds of sheep and goats, llamas, and so on), you're probably in it for the fiber. Of course, you may be raising the critters just because you like them, but why not use that fiber for something unique and truly homegrown? Read on for info on collecting an animal's fleece and getting those fibers in top shape.

Shave and a haircut: Getting the fiber off your critters

Not only does shearing allow you to collect an animal's fleece, but it also makes the animal more comfortable in hot weather. In this section, I run through the basics of shearing your fleece-producers. *Note:* You harvest the fiber from other, more exotic animals such as the yak or Angora rabbit by combing the animal, but I focus on the more-common practice of shearing.

Choosing the tools and the shearer

Electric shears make the job go much more quickly, and if you have a large herd — or even if you have more than two or three and can't afford the time needed for the hand shears — that's the way to go. Hand shears do the same job (albeit more slowly), but some people like them because the experience is more of a bonding one. You're with your animal for a longer time, and each minute makes the critter more used to your being close. Also, there's no whine of electrical shears to frighten the animal.

Shearing can be challenging in that you have to subdue the animal, get the fiber off in a useable condition, and avoid cutting the animal's skin. If you prefer not to do the shearing yourself, you can hire a professional. Each pro has his or her own unique tricks, and because the shearer does it so often, the shearer can be in and out in no time. Costs vary widely. For instance, the man who shears my goats charges $7 per goat, and the llama/alpaca shearer charges $25 per animal. However, the efficiency may be worth the cost.

Knowing when to shear or not to shear

If you let the animals go too long without shearing or combing, you're likely to end up with a fleece that's all matted and isn't useable for spinning into yarn. To get good, quality fiber, you need to shear alpacas, llamas, and sheep once a year, usually in late spring or early summer, depending on your location. Fiber goats (such as the Angora) need to be shorn twice a year, usually late winter and late summer. Table 17-1 has some notes on fiber collection and also describes the official names of various types of fiber. (Technically, the term *wool* means fiber from a sheep.)

Table 17-1	Fiber Names and Collection Notes	
Animal	*Name of the Fiber*	*Notes*
Alpacas	Alpaca fiber	Shear once a year.
Angora rabbits	Angora	Comb the rabbit out at least every other day to prevent matting.

Animal	Name of the Fiber	Notes
Goats	Mohair or cashmere	Shear twice a year.
Llamas	Llama fiber	Shear once a year.
Musk oxen	Qiviut (undercoat)	Collect the fiber by combing.
Sheep	Wool	Shear once a year.
Yaks	Down (undercoat)	Collect the fiber in the spring by combing.

Here are some tips on choosing the shearing time:

✔ **Shear the animal before the weather gets too hot.** Being under all that heavy wool can get really hot in the summertime!

✔ **Be consistent.** If you decide May is the time to shear, you should shear in May *every* year so you can get the maximum length to the fiber. One year, I scheduled the shearing for May, even though I'd done it in June the preceding year. That year, some of the animals' fleeces weren't quite long enough to be top quality because they'd grown for only 11 months.

✔ **Shear a pregnant animal just prior to her giving birth.** The birthing process is stressful on the animal, and stress reduces the quality of the fiber. For those animals who are pregnant in the springtime, have them shorn before the birth day.

✔ **Wait for a dry spell.** Wet fleece is too heavy and unworkable and can clog the shears. (Of course, weather quirks aside, the shearing season is a short one. If you're hiring a professional, you'll most likely need to make an appointment ahead of time.)

Preparing for shearing day: Getting critters a little cleaner

Animals live outside. They lie in dirt or hay or worse. That means their beautiful coats get dirty. Before you can process the fleece into something you can felt or spin with a spinning wheel, all that dirt and other stuff need to be cleaned out. Cleaning the coat before shearing is easiest (but don't despair; there are ways to do it after the shearing as well).

If you have a pasture with some good grass cover, bring the animals into that pasture and let them hang out in the grass (instead of the dirt) for a few days to a week before shearing day. If you don't have a clean place to take them, you can also spend some time hand-picking the stuff out. That's a lot of work, though.

Sheep's fleece is in a tight crimp, but the fleeces of some animals, such as the llama or alpaca, benefit from being gently blown. This air helps get some of the bigger, looser foreign matter out of the fleece. Use something like a leaf blower or Shop-Vac on a low setting.

Some people even bathe their animals before shearing. Bathing critters can definitely help you end up with a cleaner fiber — provided, of course, they don't go and roll in the mud immediately afterwards. Other people *coat* their animals — they put a little coat or sweater on the critters' backs to protect the fleece.

A good, honest fleecing: Shearing day logistics

On shearing day, everybody has to be rounded up and individually brought into the shearing area. Smaller animals such as sheep and goats are easier to round up, and because they're pretty small, shaving the fleece off and letting them go doesn't take that long. Larger animals like llamas take a bit longer, due to the surface area to shear as well as their trademark unwillingness to cooperate.

The shearing logistics are easy:

1. **Delay the morning feeding.**

   Don't feed the animal on the morning of shearing day. Wait until the shearing is all over. Shearing is a little uncomfortable anyway, and if the critter has a full stomach, that only adds to the discomfort.

2. **Catch one animal and bring the critter to the shearing area.**

   A flat, clean surface such as a concrete driveway is a good place to shear. This surface easily allows you to pick up the shorn fleeces and clean up the leftovers.

   For llamas who aren't handled regularly, shearing day may be the only time they get haltered and roped, so this can be traumatic for them. Unwilling llamas can spit, so stay out of the line of fire or be resigned to the fact that you're going to get hit.

3. **Shear off the heavy fleece (Figure 17-1) and bag it.**

   Take the *blanket* — the best part of the fleece, which goes over the animal's back. We start at the neck and make cuts close to the skin (to get the longest fiber possible), radiating out across the sides and down to the tail. The idea is to try to take this blanket off in one big piece. Put it in a garbage plastic bag.

   The other stuff off the legs and neck may or may not be useable, but you can bag that separately. Label each bag according to its contents (for example, *llama blanket, llama other,* or *alpaca blanket*).

4. **Take the newly shorn critter back to his or her pasture.**

5. **Sweep out the area after each animal has been shorn so you're ready for the next one.**

6. **Clean and oil your shears so they're ready for the next time you need to use them.**

Figure 17-1:
Shearing a
sheep.

Most raw fleece can be stored forever, provided the moths don't chew it up or other animals don't get into it. If you're going to wait awhile before you clean and process the stuff, you can leave it in the barn. However, the best way to treat fiber is to process it right away and then bring it in the house. (See the later section titled "Storing processed fiber.")

Sheep have a lot of lanolin in their fleece, giving it a greasy feel. Clean sheep fleeces right away so the lanolin doesn't harden to a point where it's difficult to clean. Lanolin can affect color and can even show up as staining where different levels of it have collected.

Processing the fiber

Now that you have some potentially beautiful fiber materials in bags, what are you going to do with them? You need to pick through the fleece, selecting only the best stuff for processing. You also have to clean it and then turn it into something you can spin or felt from.

The first step to using your fleeces is getting them processed. Processing involves a few steps: preparing and cleaning and then carding. One option is to let somebody else do the work. Just hand over the bag of freshly shorn fleece and go back later to pick up the finished product. But if you want to save a little money, you can do some preliminary cleaning yourself. This cleaning part, which I discuss in this section, is called *skirting* and *scouring*.

Skirting: Separating the best from the not-so-good

After the animal has been shorn, you have to go through the fleece and pick out the not so-great-stuff. *Skirting* is the removal of the undesirable pieces from your raw fleece. Lay the fleece out on a clean surface and hand-pick out the following:

- Contamination, including *veg* (vegetable matter — all that dried up hay stuck in the fleece), dung tags, bits of wood, wire, and assorted junk that farmers leave lying about the farm and that gets caught in the fleece when animals lie in it or rub up against it

- *Kemp* (fibers that don't have the barbs that make spinning into yarn possible); they're distinguishable because they're brittle and straight with no crimp — more like stiff hair than soft wool

- *Second cuts* (fibers that didn't get cut all the way on the first try and are thus shorter) and other fibers of inconsistent length

- Anything that's matted or felted or parts that are very course

White fleeces are very valuable because they can be dyed any color. If the tips of the hairs are a darker color, snip them off so you're left with just pure white.

You want the majority of the fibers to be the same length and the same diameter to get the best product. Part of the skirting process is to remove fiber that's inconsistent with the characteristics of the fleece. The best yarns are made from fleeces in which most of the individual fibers are the same length. Extra long fibers aren't really an issue, but the short ones are because they can poke out of the yarn, giving it a fuzzy look and a prickly feel. Also, short fibers are often coarser.

Even if all the fibers have the same length and fineness, they do need to be a certain minimum length so the fibers stay together. As the more-or-less parallel fibers come off the carder (brush or comb), each individual fiber clings to its neighbors so that they collectively form a web as wide as the carder and a few fibers thick. Longer fibers have more surface area for contact and thus form a stronger bond.

You can also rig up a *spinner,* which is kind of like a giant salad spinner, to clean off veg. The one my shearer-processor uses looks like a barrel made of a small mesh fencing material, rigged up like a meat grilling spit. As the barrel filled with dirty fleece spins, the veg matter falls out.

Scouring: Washing the fleece

Scouring is the washing of the fleece to remove dirt, *suint* (natural grease formed from dried perspiration in sheep fleeces), lanolin (a waxy substance), and other substances the animal may have embedded in the fiber.

Let the washing machine or your bathtub help with this step:

1. **Fill the washing machine (or bathtub) with hot water and liquid dish soap, about ½ cup per full tub.**

   I like to use Dawn dishwashing liquid because it chemically interacts with the grease but isn't alkaline like soap. (Dawn is what they use to wash birds who've been caught up in an oil slick, so I guess it's gentle enough to use on my fleeces.)

   Don't use normal soap for scouring; it causes the fleece to felt.

2. **Put the fleeces in light net bags.**

   I like to use those bags designed for washing delicate lingerie in the washing machine. Fill the bags loosely; if you pack the stuff too tight, there won't be enough room for the water to circulate.

3. **Lay the net bags in the water, pushing gently until they're fully wet.**

   Let them soak for about 20 minutes.

4. **Take out the bags, let them drain, and put them into the sink to drain further.**

   If you're using a washing machine, you can return the fleeces to the machine for the spin cycle. This cycle, although it spins violently, doesn't introduce agitation, so it's a good way to remove a lot of the moisture in a short time.

   Agitation (along with heat) causes the fleece to felt, rendering it unusable for spinning.

5. **Prepare and give the fleeces a second bath.**

   I find that one bath isn't quite enough, so repeat the process.

6. **Gently lay the fleeces out to dry.**

   If you have a nice, dry, sunny day, you may attempt to do this outside, but wind or curious animals can introduce some dirt back into the fleece, so I recommend drying indoors. One of those mesh sweater-drying racks is ideal. Be careful if you have cats, because a fresh fleece may look like a wonderful new bed to them! Drying takes a few hours.

Carding and such: Brushing and combing

Carding means organizing a disorganized jumble of fibers so they're more or less parallel. Carding can also remove some but not all the veg that thus far eluded the preparer. You can card by hand by using what looks sort of like a pair of big dog brushes (see Figure 17-2). Or if you have a lot of fiber to process, you can send it off to the pros.

Depending on the quality and length of the fibers, you can get several distinct end products from carding:

✔ **Clouds:** If the fibers are short and form a weak fiber-to-fiber bond that can't support its own weight so that the structure collapses, the result is a *cloud.* Spinning from a cloud is hard, so the best use for these fibers is probably needle felting (or some of the suggestions in the later "Using up all the fiber leftovers" section).

✔ **Batts:** A carder collects the fiber on a special roller, laying down layer after layer of fibers, to create a *batt.* When the roller is full (or the batt has reached the desired loft), it's broken, pulled off the roller, and left flat. People often roll batts in tissue paper to transport them.

You can spin from batts, but it's not as easy as spinning from rovings. Batts also work as stuffing; they look a lot like the quilt batts you buy in the store. They're also great for wet felting. (See the later "Locking the fibers with felting" section.)

✔ **Roving:** A *roving* is a loose rope-like preparation in which the fibers from the carder are gathered and then lightly *drafted* (pulled apart). You often find roving wound into balls or in center-pull bumps. This preparation requires the least amount of additional work before being spun by a hand spinner.

The primary criterion for roving is that the roving has to be strong enough to survive the rigors of the bump-winder so it can stay intact when tugged at during the spinning process. Due to texture, the staple length of sheep's wool that's long enough to make roving may be much shorter than the staple length needed to make a good alpaca roving.

✔ **Sliver:** A *sliver* is similar to a roving, and the two terms are often used interchangeably. Some camps say that if you add a bit of a twist is a sliver, it becomes roving. Other camps say there isn't a twist to the roving.

Figure 17-2:
Carding by
hand aligns
individual
fibers.

✔ **Rolag:** A _rolag_ is made by using a pair of hand carders (as opposed to a commercial carder) to organize the fibers. You lay tufts of fiber across the teeth of one of the carders and brush them out with the other. When the fibers are nice and straight, you peel and roll them lengthwise from the bed of the carder. All the fiber comes out of the carder, and you end up with a roll of fiber.

Cuticle fibers are what makes fibers cling to each other. Wool has a lot of rather prominent scales. Huacaya alpaca doesn't have as many scales, and the scales aren't as prominent as sheep's wool scales. Suri alpaca and mohair have even fewer scales. The interlocking of these scales — fiber-to-fiber — is the mechanism that holds roving, batts, and ultimately yarn together.

The best way to store processed fiber is to put it in clear plastic garbage bags. Moths like darkness, and the clear bags let in too much light for their taste. Store the bags in an area where mice won't chew through, or cats or dogs won't think it's some other animal's territory and thus want to pee on the fiber. I keep processed fiber in a closet in the house while it's awaiting a go at the spinning wheel.

Doing Stuff with Processed Fiber

After you process the fiber, you can do a number of things with it. You can spin it into yarn that can then be crocheted or knitted into something yummy, you can weave it into something useful and beautiful, or you can felt it into something wearable or otherwise functional. This section explores your options.

Dyeing in the wool

Natural fibers such as wool can be easily dyed (well, easily in terms that the fibers permanently take the dye, but not always so easy to do procedurally). Especially if you start out with a light color such as white or light grey, you can end up with fibers that are any number of colors from the rainbow.

You can buy one of the tons of dyes available, or you can buy books that are full of recipes for making your own natural dyes from herbs or flowers. Making natural dyes can get complicated because you need a _mordant_ (a substance that sets the dye on the wool and makes it permanent), glass pots, rubber gloves, and so on.

My favorite dye is nontoxic, extremely easy to use, and can be found on the shelves of the local grocery store: Kool-Aid. Yep, wool and Kool-Aid were made for each other. And because Kool-Aid is a food item, you can use your regular kitchen pots and utensils to do the dyeing. _Note:_ Kool-Aid works only on animal fibers, so don't try to dye cotton with it.

You can dye unprocessed fleece, processed rovings, spun yarn, or even a finished knitted sweater. In the instructions that follow, I refer to the thing-you're-dyeing as *the wool*:

1. **Pre-wet your wool by soaking it in a huge pot of warm water for about 20 minutes.**

   Wet wool absorbs the dye more evenly.

   If you're going to dye a skein of yarn that has already been wound into a ball, unwind it and rewind it into a twisted skein (a big loop). If it remains in a ball, the yarn in the center of the ball won't get enough of the dye, and the finished product won't be a uniform color.

2. **Prepare the Kool-Aid dye bath.**

   Fill a 4-quart pot about ⅔ of the way with water and pour in a package or two of powdered drink mix, depending on how rich you want the color to be (it's hard to determine what the exact outcome will be unless you do the same thing over and over, so have fun playing with different colors). Stir until it's dissolved.

3. **Add the wool and heat on the stove.**

   Add the wool to the pot, turn on the heat, and heat to a gentle boil for 5 to 10 minutes, keeping the yarn under the water by pressing down with a big spoon.

4. **Remove, rinse, and dry the wool.**

   Allow the bath to cool before taking the wool out. Rinse until the color runs clear. Squeeze out any extra water and then wrap the wool in a towel to absorb excess moisture. Let the wool dry on a sweater rack, or if it's already in yarn form, you can hang it.

   If you have something in mind for the dyed wool, dye enough at the same time so the color is uniform for the whole project.

Spinning and using yarn

If you want yarn from your own fiber, the easiest way to get it is to give the fiber to an expert who has a spinning machine. Some of the bigger operations that card your fiber can also spin it and return to you a finished product of yarn. However, many people prefer to do their own spinning. In this section, you discover some spinning tools and read about where you can pick up this craft.

Taking the wheel: Spinning implements

Processed fiber can be spun into yarn using a number of styles of spinning implements. The simplest style is a drop spindle. You don't need anything fancy to make one yourself — a dowel, a hook, and a small round disk or wooden wheel will do. A drop spindle works with the same concept as a spinning wheel — making yarn by putting a twist into the fiber — but on a much slower scale. I've heard that it takes about seven times as long to make the same amount of yarn on a spindle as it does on a wheel. But the spindles are completely portable and can be used where wheels can't, such as in the car, on a plane, or around the campfire. You can even spin while wandering around.

If you opt for a spinning wheel, you can find fold-up styles that you can pack in the car and take to spinning parties or even camping trips (which I've done, but they're not as portable and campfire friendly as the drop spindle). Some styles are not only functioning yarn-makers but also beautiful pieces of furniture to grace your living room (see Figure 17-3). And you can even get small consumer-sized electric wheels (see Figure 17-4).

Figure 17-3:
A traditional
spinning
wheel.

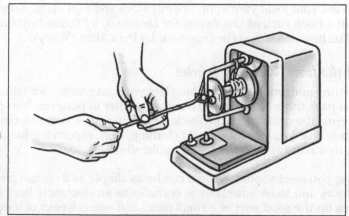

Figure 17-4:
An electric
spinning
wheel.

Try out a few different styles of wheels before you decide on one, because you may find you prefer one style over another. If you really get into this craft, you may even find the need for multiple wheels. Find a local store that sells wheels, visit fiber festivals (these typically have people who are showing off and answering questions about all things fiber), or find a spinning group in your area where somebody will let you take a spin.

Joining classes and clubs to develop your spinning technique

I suggest taking a class, having a friend show you, or at least getting video to see how to do the spinning. Figuring it out wasn't easy for me. The concept seems simple enough, but getting all the components of the process to come together to give you what you want takes a little while. Factors that come into play include the amount of twist you put into the fibers, the spinning speed, blending fibers, and so much more.

A bunch of spinning clubs are around, and chances are good you can find a group near you. Ask at the local yarn shop and join up. Not only is spinning with a group of others a fun get-together, but you can also get tips from people who really know what they're doing. *Warning:* If you sign up for a spinning class, make sure it actually has to do with fibers. Otherwise, you may be in store for an aerobic workout on a stationary bike!

A good place to start perusing all the intricacies of the craft is at the Joy of Handspinning Web site (www.joyofhandspinning.com).

Getting hooked on knitting and crocheting

Knitting or crocheting is a great way to put all that spun wool to good use. Using knitting needles or a crochet hook, you can make your own fabrics. Knitting and crocheting are great outlets for creativity because the possibilities for what you end up with are endless. You can play with stitch patterns, colors, and textures and turn out a truly unique garment, purse or fashion accessory, or something else entirely, such as kitchen curtains or teapot covers. Visit your local yarn shop or craft store and sign up for lessons, or check out a book such as *Crocheting For Dummies,* by Susan Brittain and Karen Manthey, or *Knitting For Dummies,* by Pam Allen (Wiley).

Fruit of the loom: Weaving fabrics

As in knitting and crocheting, weaving lets you make your own fabric. In this case, you pass threads under and over each other in patterns. You decide the patterns, the colors, and the thickness. Using some sort of loom, you can make fabric that you can then use in clothing, rugs, tapestries, blankets, baskets, or virtually any other fabric application.

To weave, you need a loom. A loom can be as simple as a couple pieces of two-by-fours and some dowels or as complex as an electronic floor model that takes up the good part of a small room and needs weeks of lessons to master.

Check at the local yarn store for a group to join, or visit the Handweavers Guild of America (www.weavespindye.org) for a national list of local guilds.

Locking the fibers with felting

You can make felt in a couple of ways. One method is to spin the fleece into yarn, knit or crochet it up, and then shrink it. Another is to take the unspun fiber and subject it to agitation. Whatever the method, the end product is the same: The agitation causes the individual fibers to hook onto one another, move closer together, and form a dense matt, eliminating any holes that were there before felting. Felting is a very cool thing to watch happen. This section introduces you to a few techniques.

Creating pieces of felt: Wet felting by hand

Wet felting uses hot, soapy water to achieve the effect of felting. For basic felting, gather the following supplies:

✔ A plastic tablecloth or similar covering to protect your work surface

✔ The wool to be felted

✔ A piece of netting that's a little bigger than the surface of pre-felted wool

✔ Hot, soapy water

✔ A sponge

✔ A ridged washboard

Find a smooth, clean surface to work on, such as a countertop or table, and follow these steps:

1. **Layer the fiber you want to felt on top of the plastic tablecloth.**

   If you're working with a batt, simply pull off the amount you want and lay it down. If working with roving, add the smaller pieces side by side and then add the next row so the sides overlap slightly. You want the surface to be solidly covered with the wool, with no big gaps.

   Lay out another layer on top of and across the first layer, at a 90-degree angle. Continue adding layers (four to six in total) so you provide many opportunities for the fibers to adhere to each other across layers, making a sturdy end product.

2. **Cover the wool with the piece of netting.**

3. **Dip the sponge in hot, soapy water and squeeze out the sponge so it's just damp (no water drips off); starting in the center of the wool, press the sponge into the wool, wetting it and pushing out the air.**

   Move out from the center, pressing and being sure to keep the netting in place, until all the wool is wet and soapy. Remember the following:

   - Don't sop it. If water comes up when you press down, it's too wet. With too much water, the fibers just float in the water instead of adhering to each other. Use a towel to soak up the excess.

   - The wool should lie flat. If it isn't flat — if bubbles are in the surface — press a little more water into the bubbly spot.

   At this point, you can add other strips of color for accent.

4. **Using your hands, gently rub in a circular motion, covering the entire surface of the wool.**

   Start with the netting still in place. To make sure the fibers don't try to hook to the netting, lift the netting up periodically, keeping a palm on the wool to prevent it from unhooking from its neighboring fibers.

   After you've worked the top side for a while, take the netting off and flip the fabric over. Put the netting back on top of the piece and work on the other side.

5. **When the wool is in soft felt stage, take the fabric in both hands and rub it back and forth.**

   To determine when it's in the *soft felt stage,* pinch a section and lift. If the fibers separate, it needs more rubbing (see Step 4). If it comes up as a solid piece of fabric, it's ready.

6. **Re-wet the fabric with hot water and rub it across the washboard, turning it every now and then to get it uniform.**

   Rub on all sides and at all angles. The fabric shrinks in the direction in which you rub it. The felt is done when the fabric is firm and doesn't stretch when pulled.

7. **When the fabric is done, rinse out all the soap with warm water and let the felt dry.**

   Rinse the fabric until the water runs clear. Blot out any excess water by rolling the fabric in a towel and letting it sit for a few minutes. Unroll it and lay it out again, shaping if necessary. Let it dry for around 24 hours.

Now you have a piece of fabric that you can use to make something else. For instance, you can lay out a pattern and cut it out to make a jacket or purse or use the felt as a background for a needle-felting project (see the later section on needle felting).

Felting knitted or crocheted projects: Wet felting in the washing machine

The mechanics of wet felting in the washing machine are essentially the same as they are for hand felting (see the preceding section), but you use the washing machine to do the agitation. You also have to start out with a fabric that's already sturdy, such as a knitted piece, because if you just toss some loose fleece into the machine, you'll end up with a matted glob.

The fun thing about felting is you can felt your project just a little or you can felt the snot out of it. Start out with a knitted or crocheted piece that's about one-and-a-half to three times bigger than you want it to be. Then follow these steps:

1. **Put the piece you want to felt into some sort of bag, such as a pillowcase or a lingerie washing bag.**

    During the felting process, some of the fibers come off. Keeping everything in a bag makes cleanup easier and is kinder to the machine.

2. **Set your washing machine to the lowest water level and the hottest water temperature, with a cold water rinse.**

3. **Add about ½ teaspoon of detergent, and add the bag with the fabric to the water.**

4. **Let it go through the wash cycle; stop the machine before it goes into the rinse cycle.**

    Take the piece out and check it. Most yarns take 20 to 30 minutes to felt, but your piece may be done earlier or later. If it needs more time, set the machine back to the wash cycle and let it go longer. Repeat until you've achieved the right size and density.

5. **Let the wash finish, allowing the piece to go through the rinse and then the spin cycles.**

6. **Take the piece out and shape it if necessary; let it dry.**

Entering the art gallery: Needle felting

With the *needle-felting* method, also called *needle felting,* you repeatedly poke a special, barbed needle through the wool to cause the wool's natural barbs to hook together. You can find felting needles at www.mielkesfarm.com or purchase them from other fiber craft retailers.

Besides the barbed needle, you also need a piece of foam to lay your felting material on top of so as you poke, poke, poke, the needle doesn't go down into a table (which breaks the needle) or your fingers (which really hurts).

Soap's up! Making a felted bar

When made properly, felted bars are like a washcloth and a bar of soap all in one. Because the soap doesn't just wash down the drain (more of it stays inside its casing), the soap lasts as much as three times as long as a regular bar of soap. As you use the felted bar, the wool shrinks around the bar and becomes tighter. When the soap is all gone, you can wash and dry the felt bag and fill it with catnip; pets love to play with it after a good washing.

Here's what you need:

- One bar of soap

- Enough wool to cover the soap bar completely

- A large bowl of hot water

Wrap some unfelted wool around the bar of soap (both lengthwise and widthwise) until it's completely covered. Work near a sink and have a large bowl handy. Working over the bowl, dribble some hot water on the bar while patting it lightly. Don't add too much water at first, or the wool will just fall off. Keep adding a little soapy water and patting and squeezing it lightly, shifting it around in your hands as it starts to lather up.

When the wool is completely wet, squeeze the wool-covered bar in both hands and lightly rub, dribbling on more water periodically to keep the bar nice and soapy wet. It first gets wrinkly and then foams up; then you see it starting to felt, or matt up. Keep squeezing and rubbing (don't forget to get the sides, too) until it looks like a nice, matted fabric. This takes around 15 to 20 minutes.

Dunk the bar in the bowl of water and see whether the felt casing is snug around the soap. If not, rub a little longer. If it's snug, run some cold water on it to rinse off the suds. (Cold water also serves to temporarily stop the felting process.)

You can use needle felting to do a sort of "painting" on a background. I suggest starting out with an already-felted piece as a background. (I've used it to decorate felted hats.) Although the background doesn't have to be a felted piece, it's the easiest background to use because it acts as just another layer of felt for the new dry-felted portion to dig into. Here's how the process works:

1. **Lay your background fabric over a piece of soft foam padding.**

2. **Pick a thin piece of wool you want to add to the picture and place it on the background.**

3. **Take the needle and poke several times.**

   You see the fibers joining immediately. Keep poking until the new wool piece is attached to the background.

4. **Continue until you're happy with your design.**

Your local yarn store may offer classes on needle felting, or you can see it in pictures in the tutorial on my Web site (backtobackknits.com/HowTo/NeedleFelt). If you don't have your own fiber to use, you can find needle-felting kits online that include a little bit of fiber as well as the needles, felting pad, and instructions.

You can also use needle felting to make some pretty spectacular dolls and 3-D soft sculptures. I've made some felted sheep and goats this way. You start out with a cylindrical shape as the body by rolling a few tufts of fiber and punching them with the needle so the fibers form a solid mass. For the legs, you may want to put some sort of stabilizer inside (like a piece of cording) and felt around that. Place the legs on the body and jab until they're attached. Do the same for the head and a tail.

Using up all the fiber leftovers

Some fiber, after cleaning, is too matted and can't be processed and spun. For these bits, you may be tempted to just toss them. But you can use them as stuffing for crafts (dolls, teddy bears, and so on), pillows, or even mattresses. You can also use leftover fleece as insulation in the outbuildings.

As for the fiber that's still good enough to spin but not of good enough quality or not soft enough to use in garments, you can use it to make animal blankets or household items.

Pillows and pet mattresses

Wool is sort of a magic kind of mattress material because it wicks moisture away from the body and helps you sleep a little cooler in the summer and warmer in the winter. It maintains its loft longer than any cotton fiber. Dust mites can't live in it, and it's pretty hypoallergenic. Sounds like a miracle material to me! Pillows and even mattresses are easy to make, and the end product is light and fluffy compared to its heavier shredded-rubber pillow counterpart.

Making a full-sized mattress for your own bed is probably more involved than you want to deal with, but smaller pillows or even animal beds are great candidates for a wool stuffing. You want to have several layers of fabric around the filler because some of the wool fibers can work their way through to the surface, giving the item a bearded look. Felting the outside of the pillow form may help that problem, which opens up a whole new avenue of creativity! Felt up a pillow form, decorate it with needle felting (or any number of decorations), and then stuff it with wool. A truly, completely done-by-yourself project!

To clean wool pillows and mattresses, wash them gently by hand and don't agitate. Squeeze out excess water and use a towel to blot the rest of the water. Let them air dry.

Animal bedding for shorn critters and babies

After shearing, put the loose, unspinnable fibers in the barn as bedding to help the animals' transition to life without hair. Or especially when the babies come and you need extra bedding, the leftover fiber can come in handy.

Household items

Sometimes the fiber may not be soft enough to use next to the skin. A sweater, scarf, or hat would be too itchy, and all the hard work you'd go to in order to knit it up would be in vain because you wouldn't ever wear it. So why not make something out of it that won't irritate the skin? Knitting isn't just for clothing, after all. Make a rug. Or a horse blanket. Or a wall hanging. Or a collection of stuffed animals. The possibilities are endless.

Making Creative Stuff without Fiber

You can recycle lots of farm leftovers into things decorative or useful, such as soap or picture frames. You can press wildflowers or flowers from your garden, or you can combine techniques and end up with a truly homemade work of art. This section describes some crafts designed for people who aren't necessarily fiber inclined.

Pressing flowers

A lot of the weeds on your farm produce wildflowers, and some of them are very good for pressing. You can get a fancy flower press at craft stores, or you can just use boards and weight to press the flowers. Here's the basic process:

1. **Pick your flowers.**

    Pick the flowers and press them while they're still moist, and press only the best specimens. Flowers that have brown spots or ripped petals don't look as good as the perfect ones.

    The best candidates are those whose petals are flat, such as pansies, violets, morning glories, and cosmos. You can press the entire flower whole, all the petals still connected. For flowers that have a more three-dimensional quality, such as roses and cactus flowers, press individual petals instead.

2. **Get a small piece of plywood — 12" x 12" is a good, manageable size — and lay a few pieces of plain white paper (or a piece of cardstock) over the wood.**

    The paper's flat surface between the wood and the flowers prevents the wood from pressing ruffles into the flowers.

3. **Lay out your flowers.**

    Lay your flowers on the paper, opening them and flattening them to create as much surface area as possible. Make sure the individual flowers don't touch each other. When the paper is full, lay a few more pieces of plain white paper (or cardstock) on top of that layer.

Continue making layers of flowers and paper until all your flowers are accounted for. You may want to introduce another layer of plywood between the layers if your stack gets higher than a couple of inches. Make sure the plywood layers (the pressing mechanism) are all the same size so you don't leave any flowers uncovered.

4. **Top the creation with a few layers of paper and one more piece of plywood.**

5. **Put a heavy weight on top.**

   Use a pile of heavy hardbound books, a gallon jug filled with water, or even an old beat-up baking pan filled with gravel and heavier stones.

6. **Let the flower press sit for at least a week, and then do a spot check.**

   The speed at which flowers dry varies, depending on the humidity in your air and the size, thickness, and makeup of the flowers.

7. **When the flowers are dried, store them in a zipper-lock bag.**

   Don't put too many in each bag, because you don't want them to crush each other.

Dried flowers have many uses, and they can be a wonderful accent to your homemade paper projects. You can even add dried bits of petals to the paper pulp to add color and texture. Check out one of the many books on this topic, such as *Papermaking with Garden Plants & Common Weeds,* by Helen Hiebert (Storey Publishing).

Using up twigs, leaves, and branches

You can recycle small twigs or vines, or even larger branches, into artistic and useful items. Maybe even consider making up several and taking them to an arts festival so you can make a little money on your creations. Here are a few ideas for how to use plant matter:

- If you grow grapes, twist some of the vines into wreaths. Or twist and wrap pine branches and leaves into a fresh wreath for your door, adorning it with felted balls small enough to look like berries. Wreaths are especially popular around the holidays, though they're a nice decorative element any time of year.

- Use cut branches to make miniature log houses that can serve as bird houses or even dog houses.

- Fashion branches into picture frames.

- Use corn husks to make a doll. Take a few husks, wrap them with twine, and add more husks for arms and legs. Decorate the doll with a tiny sweater or fabric dress.

Sweeping through broom-making

You can make a broom out of straw or other materials, such as broomcorn or broomstraw. First, gather straw pieces that are around 18 inches long. Then find something to use as a handle, maybe even a fairly straight branch. (Look for something that has a knob at the end — you're going to tie the straw to the handle, and with a knob, the broom will be less likely to come apart.) Then follow these steps:

1. **Tie the straw to the handle.** Lay the straw parallel to the handle so the straw ends and the end of the handle are even. Using twine, wire, or some other tying material, tie the straw to the handle at about 6 inches from the end point. Keep tying more straw on until you have enough to make a broom. (Look at a picture of other brooms to give you and idea how much you need.)

2. **Fold the straw over and tie again.** When you have sufficient straw, bend the straw at the tie so the straw is now doubled over for a few inches, but most of it is hanging off the end of the handle. Tie twine or wire around the whole thing — the handle and two layers of straw — until it's secure.

3. **Use the broom to sweep up any mess you made!**

Making candles and soap

Candles and soap are fun to make, and you can get some of the materials right from your farm. These crafty items make wonderful gifts. Either one may include beeswax, and soap can incorporate goat's milk or even honey.

Candle-making is pretty easy. All you really have to do is melt some wax, pour it into a mold, stick a wick in it, and let it dry. The soap-making process is a little more involved. To find out more about how to do these crafts, visit www.teachsoap.com or check out *Making Candles & Soaps For Dummies*, by Kelly Ewing (Wiley) http://www.dummies.com/WileyCDA/ DummiesTitle/Making-Candles-Soaps-For-Dummies.productCd-0764574086,subcat-HOME.html.

Chapter 18

Dealing with the Less Desirable Stuff

- -

In This Chapter

▶ Making do with poo

▶ Using inedible plant parts

▶ Properly disposing of chemical, construction, and machine waste

- -

*F*arm yield usually refers to good products you find on the farm, such as
fresh tomatoes, eggs, milk, or even fiber. But farm crops and animals
yield other things that aren't as obviously useful. Your first thought may be
that manure and leftover plant parts merely need to be disposed of; however,
you can recycle these byproducts and others in some creative ways.
Recycling can help keep your costs down and put useful material back in
your soil.

Of course, you'll always have something left over that you don't have the
time or resources or energy to turn into something useful. Disposing of it in a
way that's kind to the environment is the only way to go. This chapter discusses using or properly disposing of those things that you wouldn't want to
bring into the house, such as manure and piles of dead weeds. (Chapter 17
goes into the better-smelling tasks, such as using fleeces or making soap.)

Treating Animal Byproducts as Products

The primary animal byproduct you have to deal with is animal waste. You
can reuse much of the stuff that comes out of the back end of the animal in
healthy and very beneficial ways. Though it's stinky, you can use the stuff
straight up as fertilizer or add it to compost piles, providing a wonderful
(and easily attainable) source of organic materials.

In this section, I show you how to use what can be recycled as well as how to dispose of what can't. Besides manure, I also touch on methane gas and carcass disposal.

Dealing with manure when it's getting deep

Manure — perhaps a nicer, gentler name for poo — is a term that refers to either the straight stuff just as nature makes it or poo mixed with bedding materials (in other words, you don't have to separate any hay or straw that gets mixed in before using it). Either way, it can be rich in nutrients that are good for your soil.

You've heard the saying "you are what you eat," right? The same philosophy applies to manure-making. If your animals eat high-quality feed, their resulting manure is high in nitrogen, a major nutrient for plants. Lesser-quality forages and feed result in manure with a lower nitrogen content.

Choosing the manure-producing animals for fertilizer

Not all manures are created equally. Manures from animals who eat hay or alfalfa or other vegetable matter (herbivores such as cattle, llamas, goats, and so on) are good choices. They eat plant matter, which is more easily broken down by the bacteria and fungi that are responsible for turning that pile of stuff into healthy, beneficial compost.

Avoid manures from meat-eaters. These manures may contain potentially dangerous *pathogens* (disease-causing organisms that can live in meat and meat byproducts). These pathogens are not present in manures of herbivores. Manure from carnivores also tends to rot, as well as attract undesirable critters such as rats and mice.

See Table 18-1 for a listing of animals, the characteristics of the manure they produce, and how to handle the manure.

Table 18-1	Types of Manure	
Animal	*Characteristics*	*How to Use It*
Sheep, goats, and llamas	Pellet-form; healthy to use as-is	Apply it straight to the soil.
Cattle	High in nitrogen; if fresh, it can cause nitrogen burning to plants	Age or compost it before application.

Animal	Characteristics	How to Use It
Horse	Undoubtedly contains still-viable grain	Compost it first to avoid introducing weeds in the soil.
Chicken	Too high in nitrogen to be used straight; very alkaline; one of the stinkiest of manures	Age or compost it before application.
Pigs	Because pigs eat scraps that often contain meat, it may contain pathogens and heavy metals	Its use is debated (unless you feed the pigs only fruits and veggies); if you insist on using it, compost it and add a little sawdust to the mixture to help reduce heavy metals.

Aging and composting manures

Manure from ruminants such as sheep, llamas, and goats, whose poo is in pellet form, can be used straight without the need to age or compost. For instance, you can apply goat berries or llama pellets directly to your soil. (One of the chores on our farm is Poop Patrol — shoveling up the potty piles and taking them right over to the garden area.) But all other manures need some sort of treatment before you can use them.

Some manure, such as that from cows and chickens, should be aged a few weeks to a few months before using it. These manures initially have too high of a nitrogen content to be directly applied and *burn* (dry out) the plant. Aging reduces the nitrogen content as well as the number of pathogens in the manure. Simply dig the stuff into the soil and let it sit a little while before planting. It'll break down into useable nitrogen products and be a great boost to your plants.

Composting manures takes a little more work than aging them, but sending manures to the compost heap does clean up all those potty piles and is an easy way to recycle. By adding manures to other organic materials and letting the mixture cook (see Chapter 14 for more on proper composting techniques), you can turn something yucky into something that improves soil structure and helps your plants grow.

Avoiding the spread of E. coli

With recent news stories linking the bacteria *E. coli* to certain packaged vegetables and fruits, you may worry about using manure as a fertilizer. Most problems with this pathogen come about when food and waste come in direct contact, so fertilizing with manure is safe as long as you take a few precautions:

- Apply manures in spring (before the growing season) or in early fall, never during the active growing season. This gives nature enough time to kill any harmful bacteria that may be lurking in the manure.

- Use mulches and supporting structures, such as the tomato cages, to keep the fruits and vegetables off the ground to reduce the risk of their coming into contact with the manure-treated soil.

- Take care of the roots of the plants because *E. coli* can enter a plant through a damaged root. Try not to cut roots when you're working the soil. Prune out plants with broken roots.

- Wash your hands and the fruits and veggies thoroughly before eating or cooking them.

Disposing of excess manure

You somehow have to dispose of the manure from carnivores, which isn't suitable for compost. If the amount of it isn't that great, you may get away with just putting it in with the rest of the household garbage and having it hauled away to the local landfill. Or you can always till it into dirt where you aren't going to grow anything.

You may also run up against a problem where your animals produce more of the good stuff than you can collect (or want to go to the trouble of collecting) for your own use. Put a sign out in the front yard advertising manure either for sale or free for the taking. Your neighbors who garden or grow crops but don't have a lot of animals to supply their own fertilizer may jump at the chance to take some of it off your hands (figuratively speaking, of course).

Methane gas: When flatulence isn't a problem

If the price of natural gas continues to rise and reaches seriously unreasonable prices, the average Joe is going to continue to look for alternative power sources. The small farmer has a few advantages. Methane gas, that not-quite-fragrant gas that's generated by fermentation of food in an animal's stomach, can actually be harnessed and used as a fuel. Although hobby farmers usually aren't set up for or interested in using this as a major fuel source (for heating the house, for instance), you really can use this fuel to heat your barn, shop, or animal outbuildings. See Chapter 7 for information on how to actually collect this plentiful gas on a small scale.

 In California, the power company Pacific Gas and Electric (PG&E) has teamed up with local dairy farmers for what they're calling the Vintage Dairy Biogas Project, which collects liquid cow manure from the farms and stores it in a vat that covers an area as big as five football fields and 33 feet deep. They expect that amount of fuel to power 1,200 homes and plan to expand the project to produce 20 percent of PG&E's power demands.

Disposing of carcasses

Sometimes, despite all your love and care, animals die. As harsh as this is to say, that means you have to deal with the body. Sometimes that body weighs several hundred pounds, leaving you with not only an emotional or financial loss but also a big logistical problem.

If you're not set up for or aren't interested in *rendering* the animal yourself (reclaiming all parts of the animal and using those parts in some way), you have a few options. You can contact your local rendering plant, or you can dispose of the body through burial or composting (if local laws allow it).

For health purposes, wear gloves when handling carcasses. Dispose of a body as soon as possible, preferably within 24 hours.

Burying the dead

Disposing of a body yourself usually means burying it. The best burial spot is near (or as close as you can get to *under*) a tree. Bugs and bacteria take care of any flesh, leaving only the calcium from the bones and many other trace nutrients that the tree can use.

Be careful when burying, and make sure you dig a deep enough hole. Animals should be covered with at least 3 feet of soil. A large animal is typically buried in a hole around 9 feet deep. Scavengers such as raccoons and even dogs are very interested in that wonderful rot smell and may try to dig up a carcass, leaving you with a stinky mess. If the decaying carcass is too close to the surface, you may be surprised one day by seeing Fido running around with body parts clenched in his teeth. Cayenne pepper, sprinkled on the burial site, may deter the scavengers.

Bury well away from water sources to avoid contamination. Most people recommend that you bury a carcass at least 5 feet above the water table. Also avoid burying in sandy areas.

Some areas limit how much flesh per year you can bury, so make sure you stay within the limits of the law. As a hobby farmer, you're not likely to run into this problem, but keep it in mind nonetheless.

Composting bodies

You can, conceivably, compost a whole body. This may be an especially ideal option in the winter or if you have sandy soils. Composting has the added benefit of inactivating viruses and killing bacteria if the compost cooks hot enough.

Composting does involve some management, but the process isn't overly complicated, especially with smaller animals. You essentially put the body in a bin big enough to hold it, pack it in sawdust, and after a couple of years, the natural decaying process breaks down the materials to a nice, helpful compost.

Recycling critters at the rendering plant

A *rendering plant* takes waste animal products and recycles them into something useful. Hides can be processed and made into leather. Hooves and other protein products, such as hair, can become glue. (Yes, there's some truth in the old saying about horses going to the glue factory!) Any fat can be processed for use in things such as pet food. Feathers can be collected and cleaned.

Find a rendering plant by asking around or by looking up "animal recycling" in the phone book. Prices vary, as do policies. You may have to deliver the animal to the plant, or someone may come and pick up the carcass for you. Typically, that person takes only large animals, such as horses and cows, so the smaller animals usually have to be taken care of another way.

Handling Plant Byproducts

Some of the plants you grow are for the purpose of getting something tasty and healthy to eat out of the effort. Some you grow simply for the beauty they add to your landscapes. Others you grow for protection purposes, for wind breaks, or for shade. Besides the parts of the plants you intend to use — the veggies or the flowers — the plants produce corn stalks or other leaves, stems, and branches that you have to contend with. This section discusses dealing with a few of those plant parts that you can't turn into crafts or food enhancements.

Branching out in your use of trees and tree parts

Trees grow (well, that's the intention, anyway) and periodically need to be trimmed. Sometimes a tree's growth gets in the way — perhaps a low-hanging branch is blocking a roadway, or maybe a branch is getting too heavy, and intentionally cutting it off is better than having a big wind break it off. Or maybe a tree dies and you need to get rid of the dead wood, which is a fire magnet. If you have fruit trees, you need to regularly prune them so they grow better and produce more fruit.

All that trimming and pruning leaves you with branches that you have to clean up. You can burn them outright, but you have better options! Here are some ideas:

✔ Take them to a place that chips wood into mulch, or do it yourself if you have a chipper. Mulch is great for keeping soil moisture in, keeping dust down, or serving as a decorative accent around gardens. (See Chapter 14 for more about using mulches.)

✔ Add smaller branches to a compost pile to introduce beneficial carbon to the mixture (see Chapter 14).

✔ Use branches to build outdoor fires for those times when you're working outdoors in the winter.

✔ Use the smaller twigs as a fire starter in your indoor fireplace, or even take them on campouts.

Fielding corn byproducts

Besides corn's obvious use as a nutritious and delicious food, this biodegradable and renewable resource has some other amazing talents. Here's how you can use all those extra stalks, cobs, and leaves after you harvest the corn:

✔ **A heat source:** Burn the outside leaves and cobs to produce heat.

✔ **Bedding:** Bale up the leafy leftovers from your harvest and use them as straw for bedding for your animals.

✔ **Art fiber:** If you want to get skilled in the Native American arts, use corn husks to make a doll or weave a basket.

Another use for corn byproducts is *corn silage,* a product you get from fermented stalks and leaves. Silage is wonderful for cattle, especially dairy cows, because it has just the right nutrients they need to produce all that milk and it's easily digestible. Ask your cooperative extension service for the technical details of making silage.

Dealing with weeds

The best way to deal with weeds is often to simply get them off your property — put them in the garbage and let them be hauled off to the landfill.

Don't compost weeds. The seeds remain viable through the composting process, so unless you're extremely attentive to the compost pile and make sure you get the temperature and airflow right, the weeds will invade the plants you're trying to fertilize.

You can burn weeds, but that method isn't always environmentally sound. In some areas, due to too many emissions floating around in the air, burning is restricted. In California, for instance, you can never have a backyard fire. On *green burn days,* burning is permitted, but during the long stretches of *red burn days,* fires result in a hefty fine.

Managing Oils, Chemicals, Metals, and Other Waste Materials

Not everything you consider trash can legally or ethically go in a landfill. You can recycle some of it into something useful, but with other substances, you have to deal with them in an environmentally correct way.

Recycling oils and tires

With so many vanishing resources these days, recycling is essential. Instead of filling up landfills, some products can undergo reprocessing and turn into something else useful. Here are a couple of nonanimal and nonplant things you have to deal with on your farm:

- **Oils:** You can recycle cooking oils or lubricating oils, such as motor oil, into useable fuels. Take these substances to authorized centers. Motor oil can go to one of the numerous places that do quickie oil changes. As for cooking oil, take it to a local household hazardous waste drop-off center — check with your local extension service for locations. Some places may charge you a nominal fee, and other services may be free.

- **Tires:** Tires can be a very volatile source of fuel for a hard-to-extinguish fire that causes some serious pollution, which is why sending tires to the landfill is often illegal. Take tires to a recycling center to be disposed of, have them retreaded, or use them in a number of other ways — for instance, they can be put in playgrounds or used as an alternative building material.

Disposing of nonrecyclable products

Some nonrecyclable materials need to be put into proper containers and taken to your county's hazardous waste disposal centers so those who have the facilities and knowledge to deal with them can do so with the least impact to the environment. Simply dumping substances down the drain can contaminate water supplies. Solvents, heavy metals, poisons, and strong acids and bases can even damage the sewer systems. Table 18-2 lists some ideal disposal methods.

Table 18-2	Disposing of Hazardous Materials	
Product	*Disposal*	*Comments*
Paints (oil- and water-based) and paint thinners and strippers	Take them to a disposal center.	If a lot of paint is left, let somebody else use it; if you have just a little, remove the lid and allow the paint to dry out for easier disposal.
Petroleum-based products (solvents, adhesives, gasoline, diesel fuel, non-vegetable oils, some detergents)	Take them to a disposal center.	Never just put them on the ground. As little as 1 teaspoon of motor oil can contaminate thousands of gallons of water.
Insecticides and chemicals	May be handled by the fire department; call the hazardous materials people and ask about the correct means of disposal.	
Lead acid/machinery batteries (from cars, trucks, tractors)	Take them to a proper disposal center; typically, the same place that takes used oil will do.	These batteries contain lead and sulfuric acid. If you're replacing one battery with another and you don't bring in the old one, you may have to pay a core charge (maybe $7–$8) to discourage you from dumping the old one in a landfill.
Lead products (pipes, roofing materials, lead-based paint)	Take them to an authorized battery dealer.	Lead poisoning affects the nervous system, causing seizures and reduced cognitive abilities; it also affects the gastrointestinal tract, causing problems such as nausea, abdominal pain, and diarrhea.
Used building materials	Inquire at your local landfill to see what they accept. They'll let you know what to do with whatever isn't acceptable.	

If you're unsure of how to dispose of a particular waste product, call either the local landfill or one of the hazardous waste disposal centers. Or you can check the EPA Web site (www.epa.gov/epaoswer/osw) for ideas.

Part VI
The Part of Tens

The 5th Wave   By Rich Tennant

That's very nice of you, dear. But I really don't think just one beetle in the garden will do much damage.

In this part . . .

The classic *For Dummies* Part of Tens is where I answer common questions about farm life. I give you ten ideas for unique and fun activities that you probably haven't done as a city dweller, and I also dispel a few myths you may have about the lifestyle.

Chapter 19

Ten Opportunities for Community Fun

In This Chapter

▶ Visiting fairs and markets

▶ Sharing the love — and knowledge — of plants and animals

▶ Having some good old-fashioned fun

Your operation doesn't have to be all work. Life on the farm offers some unique opportunities for fun. You can try to grow the biggest pumpkin in the world and enter it in the county fair. Or you can rent out the grounds and animals for the local kids' birthday parties. What kid wouldn't love to have baby animals or animal rides as part of the celebration? You can also establish yourself as a resident expert and visit schools or community centers to give talks on that topic, sharing your knowledge and helping others in the community join in the fun. I explore these and other activities in this chapter.

Participate in County Fairs

County fairs are a lot of fun to go to as a mere spectator. Attending can be even better when you're part of the events and displays. Here are a few ways to get involved that'll lead you to the fair:

✔ **Encourage your kids to participate in youth groups.** Contact your local cooperative extension service to find a 4-H club (www.national 4-hheadquarters.gov) or a similar youth group. 4-H offers activities centering on leadership, citizenship, and life skills. The program often involves farm work, but if your kids want to branch out beyond plants and animals, they can also test their skills in cooking, photography, rocketry, and more.

✔ **Enter your produce — and homemade products — in open class competition.** Take on a project of your own and try to grow the largest veggies you can. If you're hugely successful, you may get into the *Guinness Book of World Records* for the largest pumpkin ever grown — the latest record is 1,689 pounds! Or shoot for the most unusually shaped potatoes. But that's not all: You can enter homemade items such as your prized pumpkin pie or the shawl you knitted from yarn spun from your alpaca. See how much fun this can be? Fairs even award prizes, sometimes monetary.

Regardless of the prizes at fair competitions, you can still get noticed — and that attention can lead to other things. People may come to know your name and your products, and county fair participation can lead to sales down the road. To find out what kinds of competitions are available at your fair — and to find rules and entry fees — check your state's or county's Web site, or ask your neighbors.

✔ **Volunteer at the fair.** A fair doesn't run itself, and you can find numerous opportunities to help out — such as helping set up tables and tents, unloading trailers, selling tickets, directing traffic, and so on. In doing so, you can meet and get to know your neighbors.

Sell at Farmer's Markets

Selling at a farmer's market is an easy way to get a little pocket change from the fruits (and veggies) of your labor on the farm. And it's an income opportunity that doesn't require a huge cash outlay before the fact. You can rent space for a booth for one afternoon or for the whole summer.

Besides the cash from selling at a market, you also get to enjoy the social aspect. Gathering with other farmers and sharing ideas is just fun, and socializing with the people who come to visit gives you the chance to brag a bit about your veggie varieties and growing techniques.

A lot of rules and regulations come with selling food to the general public, and each state defines its own requirements. Check local laws by visiting your state's Web site or by calling a state or county office.

Get Trained as a Master Gardener

If you live near a college, you may be able to get in on a neat program that not only helps you with your farm but also provides an avenue for giving back to the community. The cooperative extension service, which is part of your local agricultural college, offers the Master Gardener Program. Participants go through intense training and are then certified as experts in the field of gardening.

The program was set up in the 1970s as a means of sharing knowledge on horticulture and pest management between experts and the general public. Not only does the program ask for gardeners to invest time to study about the culture and maintenance of many types of plants, but it also requires that you spend a fair amount of time volunteering for your local extension service. Typical volunteer tasks include

✔ Answering telephone requests for information related to gardening

✔ Manning plant clinics or displays in community centers that help the general public with topics such as starting seeds or managing pests

✔ Volunteering in community beautification projects

Anybody who has the interest and time can go through the training to become certified as a master gardener, but living on a farm gives you a leg up because you have all that land and equipment ready to help you practice any new techniques. Plus, going through the program can help you in your growing efforts. To find a program near you, contact your cooperative extension service (locate the office by visiting www.csrees.usda.gov/Extension).

Give Lectures and Presentations

Sharing what you know with others is really rewarding. And having knowledge about running your farm (whether it's knowledge about raising crops, caring for cattle, or making cheese) gives you an opportunity to discover a lot that you can then share.

One idea is to give lectures. Schools are grateful to citizens who come in and give talks to the kids about what life in the real world is all about. You can get involved in local groups (such as the Beekeeping Society) and volunteer to do the talks as people call in and ask for them.

Ways to get noticed are numerous. You can set up a Web site for your farm, and people can find you that way. Or you can simply call schools or other organizations and volunteer yourself. They're likely to love that you volunteered, and you'll have a blast doing it.

Set Up a Farm Tour

A farm tour is an opportunity to educate as well as show off some of the things you've accomplished on your farm. Living out in a farm community may get you noticed in ways you never imagined. People may just stop by, or they may notice your Web site. Or they may see you give a talk somewhere, get interested, and want to come see the real thing. A typical visitor may be a group such as the Scouts or a home-schooled family or even a city family who has no intention of living on a farm but wants the kids to experience what it's like to do so. I even had an artist come out one day. He wanted to get some pictures of new lambs as a reference for his paintings.

Before you invite people over, do a little preparation. Fence or rope off areas that are potentially dangerous (such as a large watering trough that a child could drown in) or places you definitely don't want people to wander into unsupervised (such as a pen housing newborn animals).

If you've never led a tour before and need a trial run, take friends around to get a feel for what it'll be like. Look for ways to interject humor into your talk. Kids get a kick out of hearing that in the beehives, the girls are the ones who do all the work. And make sure to warn that some of the animals think children taste good — they like to plant sloppy kisses on anybody who shows an interest!

The actual tour starts out with a meet and greet. Sit down on the deck and have some coffee or lemonade while you take care of introductions and maybe give a general overview of what you do on the farm. Give guests an idea of what sorts of things they'll be seeing today, and maybe set some ground rules, such as staying out of roped off areas and remaining with the group.

Then you just proceed around the farm to the various buildings, gardens, and pastures, talking about what takes place there. Although you have a core set of topics you want to talk about, a lot of ideas come from whatever's going on at the time of the tour. If the animals have newborns, talk about that. If it's shearing time, you may schedule the tour on the day when the shearer is there so the kiddies (and their chaperones) can get to see how it's done. If you're making soap, you can talk about that.

Try to make the tour interactive. Here are some ideas:

- As you move into the animal enclosures, bring along some grain and let the kids stick their hands out, full of grain, and watch how they squeal with delight as they get mobbed by the sheep and goats. Explain the differences between the fibers of a meat goat and a fiber goat and let the kids feel for themselves.

- As you wander into the bird yard, the turkeys like to puff up and show off to people. Bring corn so the kids can feed the ducks and geese while you're explaining what you know about these birds and why you have them on your farm. Besides getting an education, the kids are being entertained as well!

✔ Wander into the greenhouse and show what you do in there — how you use it to process honey, how trees and other plants are started in there, how you store unprocessed wool in there, and so on.

✔ Have the kids taste berries or smell leaves and ask them whether they can guess what the plant is.

At the end of the tour, gather on the patio and answer questions. Every once in a while, you may even have a big enough group that you'll want to fire up the grill, cook some hot dogs, and visit a little longer. Then maybe give out honey or soap or eggs for the guests to take home.

What you're doing is showing what you do on a typical day and what this life that you chose is really like. Most people romanticize the lifestyle and think it's simple, but you can show them that it really is a lot of work — but if you love doing it, it doesn't matter!

Create a Petting Farm

Not only does offering your animals to the community in the form of a petting farm bring smiles to the kids' faces, but it also helps them gain a respect for animals and may even influence what they do in adulthood. One local kid who was exposed to animals at a petting farm used to spend a lot of time at my farm and eventually became a veterinarian. His little buddy went on to work in a wildlife conservation organization.

Creating a petting farm is really simple. Either designate a day when you'll have all the animal babies together in a separate holding area on your farm or take the little ones to a local event such as a fiber festival.

For babies to be part of a petting farm, they have to be old enough that their mothers will let them be away for a few hours — trying to take babies away from mothers can give you headaches (see Chapter 11 for more information on this).

Getting up close and personal with small animals (ones who are docile enough to be friendly, cute, and cuddly instead of scary and mean and potentially harmful) is something kids of all ages can't resist. However, you may want to consider getting insurance coverage to take care of any possible injuries to the kids or to the animals.

Enter Parades

Remember being a kid, when watching parades with the local marching bands and Boy Scout troops was fun because you knew somebody who was marching — or maybe you were marching yourself? Then somewhere along the years, you lost interest in parades, especially when there were no more kids in the household to get excited with.

But amazingly, when you get involved in new activities, your interests change, and sometimes surprising things trigger your childhood joy. Now that you live on a farm, you may be interested in parades again — maybe not so much the marching bands, but maybe you like to see the animals and vehicles. Now think about participating in that parade, and it's suddenly something you wait the whole year for!

Here are some ideas on how you can enter a parade:

✔ Who doesn't love to look at old machinery? That old tractor you inherited with your farm (or that you got from your farmer neighbor for a great deal) may be a big hit in the local parade!

✔ Deck out your horses in some sort of showy gear (maybe equipped for a backpacking trip or in show dress) and parade them down the street to the delight of all the kiddies — and their parents, who are now getting ideas about taking everybody for a backpacking trip with horses!

✔ How about your goats (either pack goats or hauling goats)? Put some packs on a few goats and string them together, or hitch a few to small trailers and take them for a hike where everyone can see how cool they are.

✔ Even dogs can have some fun in a parade, especially the really big breeds like the Great Pyrenees. (See Chapter 9 for more info about these dogs and why you want them on your farm.) Simply because of their size, they attract attention — and isn't that what a parade is all about?

You can find about upcoming parades by checking events listings for your area. The local paper has them, as do numerous Web sites. Maybe even the country store or someone you meet at the county fair will have good information about parades in your area. Contact the organizer and see how to get in. You may have to wait till next year, but at least you're getting in the loop.

Set Your Sights on the Small-Town Social Scene

Moving into the country is, in some ways, like stepping back in time. People tend to be friendlier. They look out for each other, and just stopping by and having a chat on the porch with some fresh lemonade is often a high point of the day. And having this as a *high point* can really be a good thing. The everyday social scene, although many city folk may find it mundane, is not only informative; it also builds a bond between neighbors that can last a lifetime.

Saturday afternoons tend to be social time. People may just stop by (lots of farm tours happen then). Or you may be out working your field and look up to find your neighbor who was also working his field is walking toward the fence for a visit. Out here, regardless of where you've been or even what you're doing on your farm, you still have a lot in common, and conversations are quick to start — as in, "I saw you were digging up your pasture yesterday. What are you planting?" or "We sure could use some more rain sometime soon" or "My apple trees don't look nearly as good as yours. What are you doing with them?"

Saturday nights in the city usually involve something like a club with drinking and dancing or watching a cultural event. In the country, there may be a barn dance or maybe a fund-raiser (including a bake sale) for an operation for one of the neighbor's kids. The activities revolve around the community and are planned (even if not deliberately) to bring the community together.

You may even find that you can't plan anything off the farm for Saturdays because people just show up and want to visit — and that's part of the fun of living in the country!

Attend or Participate in Festivals

Small towns love to have festivals. Festivals celebrate a town's heritage (Pioneer Days) or a fruit or vegetable that grows particularly well in the area (such as Melon Days or Strawberry Days). They celebrate a prominent ethnic group (such as a Greek Festival) or a local favorite drink (such as a Wine Festival). Or they celebrate a holiday or season. They go on year-round and can be great opportunities for fun and getting to know your neighbors better. Whatever the motivation for the gathering, the end result is good, clean fun.

Participating in a festival (or even organizing one!) can make it even more fun than just being an onlooker. I have a friend who every year does what she calls "Farm Days." Kids from schools far and wide come by the busload for a day of seeing what life on the farm is like. Booths teach about things such as fiber processing (complete with a sheep shearer and a display of sweaters or felted hats), hide tanning, and how wheat becomes food. It's great fun — and educational — for all involved.

Even if you can't sponsor an entire festival like that one, you can still bring your prized melons or peppers for sale or display, bring your animals for an interactive children's area, or bring your ponies or vintage tractor and offer rides at an existing festival. Find out about local festivals by watching events listings and asking around.

Experiment with Home-Grown Entertainment Opportunities

Having a farm gives you some unique opportunities for fun, such as hay rides. Load the back of your truck with some hay (loose or in bales), add several kids (of all ages!), and go for a drive. Besides a truck, you can haul a trailer with a tractor, or on a smaller scale, have a horse or cow haul a small trailer. Arrange with your local parks department for a destination, or you do it all on the farm if you have a lot of acreage. Maybe even have a bonfire and marshmallow roasting at the end of the ride.

Or how about setting up a corn maze in the fall? These can be a few simple rows or extremely complex, done by using a GPS system to cut a precise design — see the Precision Mazes Web site (precisionmazes.com/ basics) for some ideas on how to go about it.

Other ideas include having animal races, inviting a bunch of friends over for a trap shooting competition, or hosting an annual Fourth of July BBQ complete with fireworks. Let your imagination go!

Speaking of animal races, each spring, my friend's farm gets to experience a truly delightful phenomenon that he's dubbed "lamb races." Every night, just around dusk, all the new lambs just start to run around in a pack. He doesn't know what prompts the race, or who determines who gets to lead, but they just take off as one, running this way and then suddenly turning that way. If they come to an obstacle (like a log or rock), they simply jump over it. It's the funniest thing to watch. Invite your friends over for dinner and a lamb show!

Chapter 20

Ten Misconceptions about Farm Living

. . . .

In This Chapter

▶ Understanding amenities and the work load

▶ Not being alone

▶ Appreciating the farmer's coolness factor

▶ Remembering that peace doesn't necessarily mean quiet

. . . .

You may have some preconceived notions about what living in the country is like. Not all of them are grounded in reality. Some things you may think of as drawbacks (you'll become culturally deprived, or a simple life equals the Dark Ages) will actually turn into positives. And things you may look on positively at first (more abundant wildlife and less crime) can lull you into a false sense of security. In this chapter, I clear up some of these misconceptions.

Country Life Is Living in the Dark Ages

In moving to the country, you're stepping into an area where you have more space, not an area where nobody else wants to venture. Amenities are generally as plentiful as you want them to be.

For example, you don't have to use an outhouse (unless that's what you decide to do) because modern plumbing is undoubtedly available. You may not be on city water and may have to use a well, but that can be a good thing. Some well water is the best around, and you don't have to pay a monthly bill to get it! Luxuries like phone lines (sometimes even cellphones and Internet access) and cable or satellite TV go into the country, too.

Farms these days can be pretty darn high-tech. A farmer neighbor of mine has a computerized greenhouse that's wired to automatically control the temperature, watering times, and humidity. You can also install webcams — I call them "lamb cams" — in various places to check up, right from your home office, on critters in birthing pens or a quarantine area.

Failing to embrace technology in the country is a conscious decision, not a forced one.

You'll Miss Out on Cultural Activities

Chances are, a drive into a major downtown area to see the Broadway cast of *Les Misérables* at the playhouse will take a bit longer than it used to. It won't be the same as when you used to live in the city and you could walk there. But think of these activities as a special treat, and maybe even splurge on a night in a hotel.

Broadway may not come to the country, but other cultural activities do. Local schools always put on plays — sometimes they do their own versions of Broadway hits — and musical concerts and these gems even come with a far lower price tag. Some are even free!

As far as opportunities for fun, you can find loads of them, and they're very different from what you may be used to in the city. See Chapter 19 for some ideas.

You'll Lose a Sense of Community

You may think having miles between you and your nearest neighbor would decrease your sense of community, but quite the opposite is true, actually! Country folk are more gregarious and willing to just show up at your door to visit or walk right up to you outside and say hello.

You can live in a city apartment for years and not ever know your neighbors' names, let alone what they do for a living or what tickles their fancies. Not so out in the country. Out here, you and your neighbors have lots of things in common, and often everyone has the same complaint (like the lack of rain), so striking up a conversation is easy. Staying isolated is harder than quickly becoming a contributing part of the community.

This closeness also means you tend to know everybody else's business. Now don't fret — this can be a good thing, because it means you watch out for each other. With only a few houses in the space of a square mile, people tend to notice when things are out of the ordinary. If an unusual car drives into your driveway, don't be surprised if your neighbor calls to see whether you know about it and you're okay.

People also tend to jump in to lend a hand. When I first moved out to the country, the neighbors would see me fussing with the tractor and come over to ask whether they could help. One neighbor offered to show me how to plant for the best results. Unlike in the city, you don't have to take a class to get that kind of information.

And people go all out for celebrations if it means getting people together for a few hours for some socializing, catching up, and fun. It's not uncommon for a neighbor to invite everybody over to a birthday party or a luncheon after a christening or other family event. With all the friendliness and willingness to open their homes and their hearts, finding lifelong friends in many of your country neighbors is easy.

You Have to Work from Sunup to Sundown

Farms do provide enough work to keep you at it from sunup to sundown, but you don't *have to* work that hard.

You can choose to do everything and keep working until the day is done, or you can pace yourself and just do the tasks that need to be done at the moment. There'll always be that shelter that needs to be built or that tool shed that needs to be renovated or that sprinkler system that really should be updated. And if it pleases you, work on it now. But the farm probably won't fall apart if you don't.

Okay, so some things can't wait. You have to take care of broken fences, planting and harvesting of crops, and births right away. But besides these must-do chores, most things can wait. A project you don't finish today will be waiting for you tomorrow. Pacing yourself may actually be hard for some former city folk to get used to. Maybe you used to work in a high-pressure job and you had to finish certain tasks before you could start another one. For a lot of farm projects, this isn't the case.

The Air Always Smells Bad

Did you ever joke about "that good ole country air" when you smelled a not-so-pleasant odor? Well, sometimes the air can have a certain fragrance about it around a farm. How you deal with it means the difference between your happiness with the lifestyle and your thinking that your decision to move out there may not have been such a good one.

Animals have a smell, and if you keep them, you have to deal with it. Whether or not they smell bad is a judgment call. Horses smell like horses, pigs smell like pigs — they smell like they smell, and that's a fact. But the city has some pretty stinky things, too! Dogs smell. Cats smell, especially if you have to clean out the litter box. Garbage trucks smell. It all depends on what you're willing to accept.

Your perception of odor is a mindset. You can decide that the smell isn't for you and thus decide not to keep animals. Or you can look at it with the attitude that the manure is going to make your garden better, or you really like having sheep and goats in the backyard, and suddenly the smell turns into a simple animal fragrance.

People Will Call You a Hick

You're worrying about being called a hick because you've returned this favor to farmers in the past, haven't you? Get over it. Now that you live in the country, you can see how wrong and simple-minded you were back then! In fact, you'll probably hear, rather than being called a hick, that city folk tend to romanticize life on the farm and think it's simple and easy and stress free, and they'll envy you and secretly wish they could live the same way.

Or depending on what you're into, you may pick up some other names. My friend with the automated greenhouse and lamb cams often gets called a *weed* (as in, "You're such a weed") because he seems out of place. He's both a farmer (also a master gardener) and a techno-geek. Those two don't seem to go together, but he makes it work fabulously!

You may hear that you're "such a farmer," especially when you get really good at something and then share it with the community! (That, of course, is a compliment.)

Your kids may also get some unusual mileage out of living on a farm. Farm kids tend to grow up with some "cool" knowledge that their city friends don't have, such as how to use guns. A friend's daughter was invited to go trap shooting with some of the guys and she ended up out-shooting them (which apparently gave their egos a hit, because she wasn't invited again!).

You'll Be Safe from Crime

Living in the country doesn't guarantee your safety, though crime in the country tends to be different in some cases. Granted, firearms are typical on the farm, and most farmers don't hesitate to shoot if they find their property or lives are in danger. The criminally minded get to know that attacking an armed house isn't the best thing to do. People are less likely to nose around where they're not supposed to be if they know the people inside the house have a gun, know how to use it, and are not afraid to do so. However, criminals still may break in when you're not home.

Door-to-door salespeople may actually be casing your place to come back and take what they want when you're not there. Also be careful of town celebrations that the whole town turns out to attend. Sometimes the bad element just watches and waits for times like this when nobody is home.

You may see things like intentionally set barn fires or the occasional lighting of a hay patch or maybe somebody letting animals out as either a joke or as what they feel is a righteous act. But besides that, homes are typically set far enough back from the road that the approach of a strange person or vehicle is very obvious. Your neighbors are likely to notice and call you.

However, crimes of the road may still haunt you. Drag racing can occur because a lot of country roads are long and straight and not very populated. And then you may run across the possibility of DUIs. With the longer distances to travel and the lack of public transportation, people who choose to drink and drive end up spending more time on the road.

It'll Be Very Quiet

Well, the farm may be very quiet. But during certain seasons — harvest season, for example, when farm machinery runs long hours — you may run across noisy activity that can't be helped. Farmers have a short timeframe for getting certain tasks done, and for some farmers, this time of the year is when they realize the fruits of their labors. They *have* to run the equipment.

Other things that may not be very quiet include roosters, cows bellowing during calving or breeding season, and alpacas picking a fight with each other. You may hear crickets chirping or coyotes calling to each other in the distance. Depending on where you find your land, you may be able to hear things like trains coming and going or big trucks on the freeway.

But one thing that you probably won't hear is your neighbor's teenage son blasting his stereo. With houses pretty far apart in the country, hearing house noises is greatly reduced.

You (Fresh from the City) Are Smarter Than Your Country Neighbors

You may stereotype farmers as unsophisticated, but not so fast, there! You may be smarter about how to use public transportation or where the best club is, but chances are you're a mere babe when it comes to what you need to know to keep a farm alive and prosperous. You know all about city life, but your new neighbors know all about farm life. You know things that others don't, and vice versa. The good news is that pretty much everybody is willing to share and help out. And if you're open to help, their input can be very beneficial to you.

Farmers aren't typically school dropouts. Some even have doctorates or are retired lawyers! Some are master gardeners. Don't make assumptions.

You Can Treat Local Wildlife Like Pets

Yes, you're closer to nature in rural areas simply because you now hang out where the wildlife like to hang out. But trying to pull a Snow White and tame the wild animals is a bad idea. (See Chapter 4 for info on dealing with the wildlife on your farm.)

Many forms of wildlife that are beautiful from a distance can turn into wrecking balls on a well-run farm. Leave them to the wild, and don't encourage them to set up shop on your operation; otherwise, they'll become nuisances that are hard to control after you've touched their lives and patterns.

Deer, for instance, can wreak havoc on a hay pile, or they can eat the alfalfa right out of the field. You may think it's cool to have deer in your yard, but what happens when they eat up all the feed that was meant for your animals over the coming winter? Not a good situation.

Wild animals such as raccoons can carry rabies and other diseases — not a good thing to encourage coming onto your back porch where your dogs and cats or even your children hang out.

Part VII
Appendixes

In this part . . .

Here you find a couple of lists that are helpful to you, the new farmer. Appendix A lists some useful books, magazines, Web sites, and organizations that you can turn to if you want to delve further into some of the topics introduced in this book. Appendix B lists some terms you need to know and use with confidence in your new hobby.

Appendix A

Helpful Resources

You don't have to do everything alone! For a lot of the topics presented in this book, you can delve into them in much greater detail in the books, magazines, and Web sites in this appendix. Of course, you can also turn to your cooperative extension service and other local groups and services to get additional help.

Books

Lots of good books can help you with more in-depth information about specific aspects of your operations on the farm. I particularly like *For Dummies* books and the Storey Guides.

Houses, structures, and features

Building Small Barns, Sheds & Shelters, by Monte Burch (Storey)

Chicken Coops, by Judy Pangman (Storey)

Deerproofing Your Yard & Garden, by Rhonda Massingham Hart (Storey)

Fences for Pasture & Garden, by Gail Damerow (Storey)

Home Buying For Dummies, by Eric Tyson and Ray Brown (Wiley)

How to Build Animal Housing, by Carol Ekarius (Storey)

Practical Pole Building Construction: With Plans for Barns, Cabins, & Outbuildings, by Leigh Sedden (Williamson)

Root Cellaring: Natural Cold Storage of Fruits & Vegetables, by Mike and Nancy Bubel (Storey)

Animal care

Beekeeping For Dummies, by Howland Blackiston (Wiley)

Horses For Dummies, by Audrey Pavia and Janice Posnikoff (Wiley)

The Power of Positive Horse Training: Saying Yes to Your Horse, by Sarah Blanchard (Wiley)

Rabbits For Dummies, by Audrey Pavia (Wiley)

Storey's Guide to Raising . . .

> *Beef Cattle,* by Heather Smith Thomas (Storey)
>
> *Chickens,* by Gail Damerow (Storey)
>
> *Dairy Goats,* by Jerry Belanger (Storey)
>
> *Meat Goats,* by Maggie Sayer (Storey)
>
> *Llamas,* by Gale Birutta (Storey)
>
> *Pigs,* by Kelly Klober (Storey)
>
> *Sheep,* by Paula Simmons and Carol Ekarius (Storey)

Gardening and crops

Burpee: The Complete Vegetable & Herb Gardener: A Guide To Growing Your Garden Organically, by Karen Davis Cutler (Wiley)

Burpee Complete Gardener: A Comprehensive, Up-To-Date, Fully Illustrated Reference for Gardeners at All Levels, by Allan Armitage, Maureen Heffernan, Chela Kleiber, and Holly H. Shimizu (Wiley)

Carrots Love Tomatoes: Secrets of Companion Planting for Successful Gardening, by Louise Riotte (Storey)

Gardening Basics For Dummies, by Steven A. Frowine and the National Gardening Association (Wiley)

Let it Rot! The Gardener's Guide to Composting, by Stu Campbell (Storey)

Organic Gardening For Dummies, by Ann Whitman and the National Gardening Association (Wiley)

Enjoying and selling the fruits of your labor

Canning & Preserving For Dummies, by Karen Ward (Wiley)

Crocheting For Dummies, by Susan Brittain and Karen Manthey (Wiley)

The Joy of Spinning, by Marilyn Kluger (Owl Books)

Knitting For Dummies, by Pam Allen (Wiley)

Making Candles & Soap For Dummies, by Kelly Ewing (Wiley)

Not Your Mama's Felting: The Cool and Creative Way to Get It Together, by Amy Swenson (Wiley)

Small Business For Dummies, by Eric Tyson and Jim Schnell (Wiley)

Spinning in the Old Way: How (and Why) to Make Your Own Yarn with A High-Whorl Handspindle, by Priscilla A. Gibson-Roberts (Nomad Press)

Teach Yourself VISUALLY Handspinning, by Judith MacKenzie McCuin (Wiley)

Magazines

You can visit www.world-newspapers.com/farming.html for a list of some more-specialized magazines catering to the farmer and agricultural enthusiast, or check out some of the publications I list here.

Hobby Farms: www.hobbyfarms.com

Farmers Hot Line: www.agdeal.com/publications/farmershotline/regional.cfm

Small Farmer's Journal: www.smallfarmersjournal.com

Small Farms: www.smallfarms.net/sfm.htm

Small Farm Today: www.smallfarmtoday.com

Spin-Off, Interweave Knits, Interweave Crochet, and other fiber magazines: www.interweave.com/magazines

Web Sites

The Internet puts a world of information literally at your fingertips. You can find loads of info on everything from local laws to see-it-in-pictures tutorials for just about anything you want to know. Here are some Web sites I find helpful.

General

AgWeb agricultural homepage: All about the agricultural industry: weather reports, market trends, equipment classified ads, discussion boards, and the like (www.agweb.com)

Computers on the Farm convention: An annual convention in Missouri where you can find out what you can computerize and what's new (agebb.missouri.edu/cotf)

Cooperative Extension Service: A list from the USDA on local cooperative extension offices (www.csrees.usda.gov/Extension)

National Agricultural Library (NAL): A massive collection of resources on crops, livestock, agricultural laws, marketing, natural resources, rural community development, and more (www.nal.usda.gov)

National Gardening Association: Expert gardening advice from the NGA (www.garden.org)

Noble Foundation Agricultural Information Index: Tons of articles on farm-related topics, such as "Answer Key Questions Before Choosing a Water Pump System," "Avoiding Plant Diseases," and "Fall is Just Around the Corner — Are You Ready?" (www.noble.org/Ag/Research)

Small Farm Center: A small-farm resource from the University of California (www.sfc.ucdavis.edu)

State laws: Find a lot of the rules and regulations that farmers come up against on your state's Web site; search for guidelines on zoning, wildlife handling, and so on (usually www.<your state>.gov)

United States Department of Agriculture (USDA): Good information on official stuff such as laws and regulations, current goings-on such as studies and trade news, and articles on all things agriculture (www.usda.gov)

USA Gardener: Information about and help with growing fruits, veggies, flowers and herbs (www.usagardener.com)

Equipment and supplies

Intermountain Farmers Association (IFA) Country Store: If your local country store doesn't carry something, you may be able to find it here (www. ifacountrystore.com)

Craigslist classifieds: Craigslist is a national want-ad Web site with many local hubs; buy and sell farm equipment, services, and so on (www.craigslist.org)

Duct tape uses: Here you can find creative and helpful uses for duct tape, submitted by individuals; look at how others have used duct tape for vehicles, home repair, gardening, and even livestock (www.duckproducts.com/creative)

WD-40 uses: This site lists the amazing things WD-40 can do, submitted by contractors, professionals in the home repair service, and homeowners (www.wd40jobsite.com)

Alternative power, the buy-local movement, and the environment

Alternative Farming Systems Information Center: The National Agricultural Library's info on alternative energy, organic farming, grazing systems, ecological pest management, and more (afsic.nal.usda.gov)

Build It Solar: Free information on renewable energy for do-it-yourselfers (www.builditsolar.com)

Collaborative Biodiesel Tutorial: How to make biodiesel, from individuals who make it (www.biodieselcommunity.org)

Database of State Incentives for Renewables and Efficiency: A state-by-state list of grants, rebate programs, and other energy-saving incentives (www.dsireusa.org)

Environmental Protection Agency (EPA): Answers to questions about what you can do to help protect the environment (www.epa.gov)

How to Go Organic: The Organic Trade Association's extensive resources for those interested in going organic (www.howtogoorganic.com)

Local Harvest (buy local): A site connecting farmers with customers who want to buy farm products from their local communities (www.local harvest.org)

National Sustainable Agriculture Information Service: Info on biodiesel, ethanol, solar power, geothermal energy, wind power, and other alternative energy sources on farms (attra.ncat.org/energy.php)

National Renewable Energy Laboratory: A renewable energy site for farmers and ranchers (www.nrel.gov/learning/farmers_ranchers.html)

Off-Grid: Issues and information for those wanting to live off the power grid in a sustainable environment (www.off-grid.net)

Organic gardening: A collection of short articles on organic gardening (www.organicgardening.com)

Renewable energy and the farm bill: Notes on the Farm Bill and ideas on applying for renewable energy and efficiency programs (www.rurdev.usda.gov/rbs/farmbill)

Animals and animal feed

Breeds of cattle, goats, and other livestock: www.ansi.okstate.edu/breeds

Llama care: www.llamacrossing.com and www.llamas-information.com

Llama and alpaca training (CAMELIDDynamics): www.camelidynamics.com

Pot-bellied pigs: www.pigs4ever.com

Pasture management: www.livestocktrail.uiuc.edu/pasturenet

Making corn silage: extension.missouri.edu/explore/agguides/crops/g04590.htm

Plants and pest control

Hort Corner: A site containing large lists of fruit and vegetable crops and how to grow and maintain them, pest control recommendations, and more (www.urbanext.uiuc.edu/hort)

Weed management: All you ever wanted to know about taming the weeds in your backyard jungle (edis.ifas.ufl.edu/TOPIC_GUIDE_Weed_Management_Guide)

Biological weed control: Ideas for biological weed control — which critters are the most effective on the types of weeds you're contending with (www.integratedweedcontrol.com)

Corn gluten meal (herbicide): Info on using a byproduct of a corn-refining process as a natural herbicide (www.hort.iastate.edu/gluten)

Homemade pest control: Tons of clever home remedy ideas for pest control (www.herc.org/maketheconnection/docs/Volume3.pdf)

Arbico Organics (organic pest control): Ideas and products for helping you deal with pests organically (www.arbico-organics.com/organic-pest-control.html)

Crop rotation recommendations: A pretty large list of ways to use crop rotation to minimize plant disease (mtvernon.wsu.edu/plant_pathology/veg_rotations.htm)

Enjoying the fruits of your labors

Pick-your-own farms: Picking tips, canning directions and supplies, and a list of U-Pick farms around the country; if you decide to run one of these farms, be sure to register here (www.pickyourown.org)

National Center for Home Food Preservation: Techniques, recipes, and so on for preserving foods (www.uga.edu/nchfp)

Canning foods: Lots of great techniques for canning and preserving fresh foods, from the makers of Ball and Kerr jars (www.freshpreserving.com)

Smoking meat: All sorts of tips and recipes for smoking meats (www.smoking-meat.com)

Hand-spinning: A great resource for the hand-spinner — tips and techniques (even videos showing how things are done), articles, and products (www.joyofhandspinning.com)

Natural disasters and damage

Emergency survival: A great Web site with ideas on what to keep on hand and what to do in the case of an emergency (www.72hours.org)

Wildfire protection: More good information on protecting your house and land from fires (csfs.colostate.edu/protecthomeandforest.htm)

Windbreaks: A guide to protecting your property from wind damage (www.extension.iastate.edu/publications/pm1716.pdf)

Federal Emergency Management Agency (FEMA): Lots of ideas for preparing for and riding out natural disasters, including information on preparing and handling livestock in a crisis (www.fema.gov)

Organizations

Alpaca Owners & Breeders Association
5000 Linbar Dr., Suite 297
Nashville, TN 37211
Phone 615-834-4195
E-mail info@aobamail.com
Web site www.alpacainfo.com

National Bison Association
8690 Wolff Ct., Suite 200
Westminster, CO 80031
Phone 303-292-2833
Web site www.bisoncentral.com

American Dairy Goat Association
209 West Main St.
Spindale, NC 28160
Phone 828-286-3801
E-mail info@adga.org
Web site www.adga.org

American Meat Goat Association
P.O. Box 676
Sonora, TX 76950
Phone 325-387-6100
Web site www.meatgoats.com

American Quarter Horse Association
P.O. Box 200
Amarillo, Tx 79168
Phone 806-376-4811
Web site www.aqha.com

North American Potbellied Pig Association
15525 E. Via Del Palo
Gilbert, Arizona 85298
Phone 480-266-8755
Web site www.petpigs.com

American Sheep Industry Association
9785 Maroon Circle, Suite 360
Centennial, CO 80112
Phone 303-771-3500
E-mail info@sheepusa.org
Web site www.sheepusa.org

Handweavers Guild of America
1255 Buford Highway, Suite 211
Suwanee, GA 30024
Phone 678-730-0010
E-mail hga@weavespindye.org
Web site www.weavespindye.org

Appendix B

Glossary

● ●

4-H: An organization designed to help kids get an education in and have fun with agricultural and other life skills; the h's stand for head, heart, hands, and health

acidity (of soil): How low the pH is (a measure of hydrogen ions); soil is acidic if it has a pH below 7

aerate: To poke the soil to allow airflow and to increase water penetration

aerobic: With oxygen (if oxygen is involved in composting, it's an aerobic process)

alkalinity (of soil): How high the pH is (a measure of hydrogen ions); soil is alkaline if it has a pH above 7

all-natural (food, farming): Farming using practices and materials that are all natural (that is, no chemical pesticides, no genetic engineering, and so on)

anaerobic: Without oxygen (if oxygen isn't involved in composting, it's an anaerobic process)

angora: The fiber produced by an Angora rabbit

balage: Alfalfa or grass that's too wet to be stored for hay feed, so it's cut and baled and stored in plastic wrappers to ferment

batt: A form of processed fleece that's in the form and shape of a sheet

beef cattle: Cattle raised for their meat

berm: A wall or mound of earth used to act as a barrier, as in separating different planting beds

billy: An adult male goat

biodiesel: An alternative fuel made from plant oils that you can use in diesel engines

biological control: A pest control method that uses plants or animals rather than chemicals to control pests

biomass: Organic matter available on a renewable basis

blanket (fleece): The prime part of the fleece, the blanket is the part that covers the animal's back from the neck to the tail

boar: An adult male pig

Bt *(Bacillus thuringiensis):* A naturally occurring soil bacteria that rots out the innards of certain bugs

buck: An adult male goat or rabbit

camelid: A member of a family of cud-chewing animals who have three-chambered stomachs; animals in this family include llamas, alpacas, guanacos, and vicuñas

carding: Organizing a jumble of fibers so they're more or less parallel; this process turns fleece into a spinnable product

castrated: Describes male animals whose sexual equipment has been snipped and who can no longer produce babies

cattle: A collective term referring to a group of domesticated bovine (cow-related) animals; a female is called a *cow,* an intact male is a *bull,* a castrated male is a *steer,* and the young is a *calf*

chevon: The meat from goats

cistern: A large water-holding tank, typically used in areas where fresh water is scarce

cloud: A form of processed fleece that's not of a high enough quality to spin from but can be used in felting, stuffing, insulation, and so on

cock: An adult male chicken; a rooster

cockerel: A young male chicken

cole crop: *See* crucifer

colt: A young male horse

compost: A soil treatment resulting from the decomposition of organic matter

cooperative extension: A service set up by the local agricultural college to share knowledge between the experts and the general public

cover crop: A fast-growing plant that spreads over the ground to protect and help improve the soil; cover crops are often planted during the off-season to benefit the plants that are regularly planted in a space

crop rotation: The practice of planting different things in the same spot as a measure of replenishing soil and curbing pests and disease

crucifer: A plant in the mustard family, including cabbage, turnip, collard, mustard, and radish

cultivar: A contraction of *cultivated variety,* meaning a plant variety (subspecies) specially bred for a feature that distinguishes it from the variety it was selected from

curing (meat): Adding a solution of salt to aid in meat preservation

dairy cattle: Cattle raised for their milk

dairy goats: Goats raised for their milk

diatomaceous earth: A nontoxic insecticide made up of the crushed shells of *diatoms* (a type of single-celled organism); it feels like talcum powder to you, but to a bug, it's like walking over crushed glass

dicot: Short for *dicotyledon,* a plant whose seed grows by producing a root that pushes the seed up out of the ground

doe: An adult female goat or rabbit

draft cattle: Cattle used to help with work on the farm (pulling plows, for instance)

dress out (an animal): To kill and prepare an animal for consumption

dry felting: *See* needle felting

E. coli: A type of bacteria often found in animal waste materials that can cause serious health problems if it comes in contact with food

ewe: An adult female sheep

farrier: A blacksmith specializing in shoeing horses and trimming hooves

farrow: The process of giving birth to a litter of piglets; a farrowing pen, farrowing house, and farrowing crate are places where a sow farrows

felt: Fusing the individual fibers of wool together, resulting in a dense, matted fabric

ferment: The process of pickling through the decomposition of carbohydrates

fiber: A natural filament (the hair from animals such as llamas or sheep) that you can twist and spin into yarn

filly: A young female horse

fixing nitrogen: The ability of some organisms (such as bacteria that live on legumes) to take nitrogen in its simplest form out of the atmosphere and convert into soil nutrients

fleece: The fiber sheared from a fiber animal (such as a sheep, alpaca, or Angora goat)

foal: An infant horse

gander: An adult male goose

gelded: A synonym for *castrated;* this term commonly refers to horses

gelding: A castrated male

geothermal energy: Energy that's generated by converting hot water or steam from deep beneath the Earth's surface into electricity; may also refer to heat energy derived from soil several feet below the surface, which usually remains a constant 50–60°F

grandfather clause: An exemption clause attached to laws (such as zoning laws) that allows people already in operation to continue business as usual even if new laws prohibit those operations going forward

gray water: Undrinkable water that isn't as icky as sewer water and is suitable for laundry, baths, dishes, and so on

Green Belt designation: Based on a law passed in 1974, a label that reduces property taxes of land actively involved in agricultural operations

greenhouse: A building where plants are raised or protected

greenspace: A section of land that won't be developed — it'll remain green with grass or brown with sand or whatever color nature gave it

haylage: A mixture of hay and grass that's put into a silo or pit and allowed to ferment

hen: An adult female bird

herbicide: A substance used to control weeds

hogget: A year-old sheep

homogenize (milk): A process by which the size of fat globules in milk are reduced so they remain suspended in the milk instead of rising to the top

Huacaya: A type of alpaca, the Huacaya (pronounced wah-*kie*-ya) produces a sheep's wool–like fiber that's dense and soft; the Huacaya alpaca looks like a small, fluffy llama

hutch: A rabbit enclosure

hydroponics: The science of growing plants in mineral solutions or liquid instead of in soil

hydropower: Energy produced by harnessing power from moving water

immersible: Able to be completely covered in water without harm

intact animals: Animals whose sexual equipment is still functioning (opposite of gelded, castrated, spayed, or neutered)

kid: A young goat

kitten: A young rabbit or cat

lamb: A young sheep; also refers to the meat from a young sheep

leaching: The removal of soil nutrients as they dissolve in the water that goes through the soil

lean-to: A three-sided structure with a slanted roof

legume: A plant that lives in a mutually beneficial relationship with bacteria that fix nitrogen from the atmosphere, converting nitrogen into a form the plant can use

liquefied petroleum gas (LPG): The liquid form of propane that's used in home heating

mare: An adult female horse

master gardener: A person who has gone through the Master Gardener Program, a program sponsored by extension services of local universities; the program provides hours of training in gardening, and in return, the graduates give back to the community

meat goats: Goats raised for their meat

mohair: The fiber produced by an Angora goat

monocot: Short for *monocotyledon,* a plant whose seed splits and throws up a shoot from within it, keeping the seed underground

mulch: Materials placed over soil to keep moisture in, to keep the sun off, or to control weeds

mutton: The meat from older sheep

nanny: An adult female goat

needle felting: Using a barbed needle to interlock wool fibers and thus felt it

neutered: A synonym for *castrated,* typically referring to dogs and cats

nightshade: A family of flowering plants that produce chemicals known as alkaloids, many of which are toxic; some of the plant parts are edible

NPK (fertilizer): Nitrogen, phosphorus, and potassium, the three main plant nutrients found in most fertilizers

off-grid: *Grid* refers to existing utility lines, so being *off-grid* means you're not hooking into existing lines

organic (food, farming): Using no chemical pesticides, herbicides, or synthetic fertilizers in an agricultural operation

organic matter: Any material that originated as a living organism, such as decaying plant or animal remains

organic pesticide: A pesticide made from a natural product that has undergone very little processing

pasteurize: The process of heating milk for a short time to kill off most of the harmful pathogens it contains

permaculture: The design of sustainable human habitats; for everything you take out of the ground, you put something back in

permit: An authorization to build or to engage in certain activities; you need a permit to build a house or to make ethanol (ethyl alcohol)

pH: A measure of hydrogen ions, indicating the acidity or alkalinity of a solution (or in the scope of this book, of soils)

piglet: A young pig

pole barn: A structure whose main supporting beams are poles or beams

pullet: A non-egg-laying female chicken under 1 year of age

qiviut: The warm, lightweight fiber of the musk ox

quarantine: To isolate or separate an animal from others, usually due to an illness that you don't want that animal to pass on to the rest of the herd

quonset hut: A rounded building, typically made of steel; picture a giant pipe, cut in half, leaving a shape with a flat bottom and a domed top

raised bed: A gardening area where the soil has been elevated above ground level, typically used in areas where drainage isn't very good

raw fleece: Just-shorn fleece, not yet cleaned or otherwise processed

rolag: A form of processed fiber that comes off a hand carder

root cellar: A usually underground room that provides a cool and humid environment for storing and prolonging the life of fruits and vegetables

rotational grazing: The practice of giving pastures a regrowing period (keeping animals off it) between grazing periods as a measure of replenishing pasture grass and preserving soil

roving: Processed fiber in the form of a loose rope-like preparation

ruminant: A hoofed animal, such as a cow or camelid, who has a three- or four-chambered stomach and chews cud

scouring: Washing fleece to remove dirt, suint (grease), lanolin, and dung

shearing: Removing the fleece from hairy animals by cutting it with electric shears or hand shears; shearing is done once or twice a year, depending on the animal

shears: Scissors-like or electric-razor-like cutters used to take the fleece off of animals

silage: A mixture of crops (such as corn, grass, or oats) that's cut up into small pieces and put into a silo or pit and allowed to ferment

skirting: The process of removing all the undesirable pieces from raw fleece

sliver: A rope-like form of processed fiber; the term is often used interchangeably with *roving*

smoking (meats): Using smoke from hardwoods to preserve and dry meats

sow: An adult female pig

stables: Small partitions in a shelter that house individual animals

stallion: An adult male horse

stock: Short for *livestock*

subsistence: Having to do with the minimum, basic needs for living — obtaining food and shelter

suint: Natural grease formed from dried perspiration found in the fleece of sheep

Suri: A type of alpaca that produces a long and silky fiber; the Suri looks like it's wearing dreadlocks

sustainable: Providing for the needs of the present without compromising the ability of future generations to meet their needs

tack: The gear you use for working horses, including bits, halters, lead ropes, and saddles

tom: An adult male turkey

variety (plant): A plant subspecies; one of the forms a specific plant takes (you can choose from many varieties of tomatoes, roses, squashes, and so on)

veg: Short for *vegetable matter;* the extra hay, leaves, and so on that clings to an animal's fiber that you need to pick out before processing and spinning

warp: Lengthwise yarns (or the background support) into which the weft wool (horizontal yarn) is woven

weft: The yarn that's drawn over and under the warp yarns to produce a patterned woven fabric

wet felting: Using hot water with a bit of soap and agitation to felt wool

wethered: A synonym for *castrated,* most commonly referring to sheep and goats

yak: A long-haired bovine that's native to Tibet and throughout the Himalayan region of South Central Asia, as well as in Mongolia

Index

toolboxes, 79
ToolCats, 91, 95
tools
  animal care, 88–90
  electrical, 83–84
  fencing, 81, 93
  gardening, 86–87, 91, 98
  general, 79–82
  maintenance, 91, 92–93
  metalwork, 84
  outdoor, 85–86
  plumbing, 84
  safety, 57, 91, 98
  selecting, 77–79
  snow-removal, 78, 93
  storing, 74
  vehicles, 94–96
tornadoes, 75–76
tours, farm, 321–323
town, travel to, 24–25
tractors, 57, 78, 92, 94–95, 114
traffic, rural, 22. *See also* roads
trailers, 96
traps, 61, 200, 253
trash disposal, 32
traveling plant varieties, 221
treats, animal, 185–186
trees, 58–59, 125–128, 237, 249, 256, 312–313
trimmer sharpeners, 89
trimmers, 87
tropical plants, growing, 257
trucks, pickup, 94
TTEAM (Tellington Touch Every Animal Method), 155
tubers, 232
turbines, wind, 113
turkeys, 54, 140, 148
turnips, 35
twine, bailing, 82

• U •

ultrasonic devices for pest control, 61
umbilical hernia, 214
undulant fever (UF), 161, 202

United States Department of Agriculture (USDA) resources, 221, 338
United Vet Equine, 198
University of Washington Extension crop rotation information, 224
unsophisticated, farmers perceived as, 330, 332
U-Pick produce, 271–272, 341
urea, 248
urine as pest control, 61
USA Gardener Web site, 338
USDA (United States Department of Agriculture) resources, 221, 338
utility knives, 86
utility trailers, 96

• V •

vacations, 12, 16, 21, 25
vaccinations, 201–203
veg in fiber, 292, 293
vegetables, 35, 76, 186, 225, 268–269, 277–280
vegetative filter strips, 128, 132
venison, 162
vermin, 13, 21, 53, 90, 137, 253
vernal ponds, 117–118
verticillium wilt, 238
veterinarians, 142, 180, 193, 203, 208–209
Victory Seed Company frost date chart, 223
vicuña, 149
Vidalia onions, 15, 233
Vietnamese Potbelly pigs, 142, 340, 342
Vintage Dairy Biogas Project, 310
vitamin supplements, 207
voltmeters, 83

• W •

Walla Walla onions, 15, 233
walnut trees, black, 237, 249
waste disposal, 125–126, 311–315. *See also* manure
watchdogs, 136–137